TEXTBOOK OF PTERIDOLOGY

TEXTBOOK
OF
PTERIDOLOGY

Dr. Pooja
Dept. of Botany
R.C.C. College
Ghaziabad (U.P.)
(India)

DISCOVERY PUBLISHING HOUSE PVT. LTD.
NEW DELHI-110 002

First Published - 2010

Reprinted - 2015

ISBN: 978-81-8356-581-3

Textbook of Pteridology

Published by:

DISCOVERY PUBLISHING HOUSE PVT. LTD.

4383/4B, Ansari Road, Darya Ganj

New Delhi-110 002 (India)

Phone: +91-11-23279245, 43596064-65

Fax: +91-11-23253475

E-mail: discoverypublishinghouse@gmail.com

sales@discoverypublishinggroup.com

web: www.discoverypublishinggroup.com

Printed at:

Infinity Imaging Systems

Delhi

Preface

Ferns have a big advantage over the mosses in their vascular tissue. They can grow taller, and can exist in more diverse environments. This is a trend that will continue in evolution, eventually leading to the rise of such large sporophyte generations as the great sequoia trees. But if ferns are so much more fit for survival, why are there still mosses? And if a larger sporophyte generation is more fit, why haven't sequoias become dominant enough to eliminate the ferns? While there are clear benefits to a larger sporophyte generation, in some recurring natural situations, natural selection favors mosses over ferns or ferns over trees. Spores are better at spreading by wind than many seeds are, for instance. So while in the long term, the protection of a seed allows seed plants to be dominant on the planet, in many situations the lightness and transport of a spore is still efficient in spreading ferns.

There are two gena that have showed imense diversity in the divsion Pteridophyta. Selaginella and Equisetum have been identified to be the only gena that are heterosporous.

Having a large sporophyte allows ferns to produce many more spores than a moss could—recall that each sporophyte on a moss only carried one sporangia. Producing many more propagules increased fern presence and dominance. Besides having a larger sporophyte generation, ferns have many important adaptations that increase their capabilities above the mosses. Ferns have roots, which, unlike moss rhizoids, not only anchor, but take up nutrients. Ferns are vascular plants, with lignified vascular tissues. These allow active water transport. That water transport along with the strength of the ligified cells allow ferns to be much larger than their moss ancestors. At one point, ferns and fern trees were the most advanced plant life, and grew even larger than ferns today do, with great size and variety of ferns. There were no flowering plants in the early cretaceous- the first forests of the dinosaurs were composed of fern trees.

Plants are the foundation of all life on earth, without which we cannot survive. India is a mega biodiversity country with about 13,000 species of vascular plants including about 1000 species of ferns and fern allies.

– Author

CONTENTS

1
Introduction

Pteridology is the study of ferns—plants classified in the Division Pterophyta (or Filicophyta). Ferns do not have seeds the way trees and flowering plants do. Rather, they have spores the way mosses do. The haploid spores grow small haploid organisms, which then undergo fertilization and grow the diploid fern plant directly out of the haploid gametophyte, similar to the sporophyte stalk growing out of the moss. The larger part, what we think of as the fern, is the sporophyte. The gametophyte is a small green prothallus that the sporophyte grows out of. Ferns are still tied to an aquatic environment, in that once a spore grows into a prothallus, there must be moisture enough for the egg in the prothallus to be fertilized by swimming, flagellated fern sperm.

Having a large sporophyte allows ferns to produce many more spores than a moss could- recall that each sporophyte on a moss only carried one sporangia. Producing many more propagules increased fern presence and dominance. Besides having a larger sporophyte generation, ferns have many important adaptations that increase their capabilities above the mosses. Ferns have roots, which, unlike moss rhizoids, not only anchor, but take up nutrients. Ferns are vascular plants, with lignified vascular tissues. These allow active water transport. That water transport along with the strength of the ligified cells allow ferns to be much larger than their moss ancestors. At one point, ferns and fern trees were the most advanced plant life, and grew even larger than ferns today do, with great size and variety of ferns. There were no flowering plants in the early cretaceous- the first forests of the dinosaurs were composed of fern trees.

Evolution and Ferns

Ferns have a big advantage over the mosses in their vascular tissue. They can grow taller, and can exist in more diverse environments. This is a trend that will continue in evolution, eventually leading to the rise of such large sporophyte generations as the great sequoia trees. But if ferns are so much more fit for survival, why are there still mosses? And if a larger sporophyte generation is more fit, why haven't sequoias become dominant enough to eliminate the ferns? While there are clear benefits to a larger sporophyte generation, in some recurring natural situations, natural selection favors mosses over ferns or ferns over trees. Spores are better at spreading by wind than many seeds are, for instance. So while in the long term, the protection of a seed allows seed plants to be dominant on the planet, in many situations the lightness and transport of a spore is still efficient in spreading ferns.

There are two gena that have showed imense diversity in the divsion Pteridophyta.Selaginella and Equisetum have been identified to be the only gena that are heterosporous.

Identification of rare and endangered plant species is the first requirement for any conservation programme. The IUCN guideline is the only available method to identify the rare and endangered species and it requires vast data on the wild population of the target species. None of the biological characters, which are playing main role in the survival and distribution of several species, is used in IUCN guideline. In the meantime there are several difficulties in following IUCN guideline, particularly the non availability of complete field data. Moreover, the same guideline can not be used for all the groups of species in equal importance. The vascular cryptogams, pteridophytes, are also an important component of any mountainous flora and they have also to be conserved in nature. As they are the primitive vascular plants on the earth, they are getting depleted in the flora due to various reasons and it is the right time to identify the rare and endangered pteridophytes to conserve them. By considering various difficulties of IUCN method for the identification rare and endangered pteridophytes, a very simple method has been adopted by using just four criteria and this method can be applied to Pteridophytes from any region of the world.

Plants are the foundation of all life on earth, without which we cannot survive. India is a mega biodiversity country with about 13,000 species of vascular plants including about 1000 species of ferns and fern allies. It is our response to conserve them for the future generations. To conserve the organism, whether plant or animal, the first step is to identify the rare and endangered organisms in a given geographical area or country. With the aim to conserve the worlds' flora, the IUCN (International Union for Conservation of Nature) has made special efforts to identify the rare and endangered vascular plants. The information contained in the 1997 IUCN Red List of Threatened plants is useful for conservationists. Worldwide, 12.5 per cent of the world's vascular plants are threatened with extinction, and it has been shown that in areas with more complete coverage, even higher numbers of threatened species (20 per cent to over 40 per cent on some islands) are being recorded. There are an estimated 270,000 known species of vascular plants, which include ferns, fern allies, gymnosperms and flowering plants. Of the species assessed by IUCN, 33,798 species, or at least 12.5 percent of all known vascular plants, are threatened with extinction on a global level. These plants are found in 369 families, and are scattered throughout 200 countries around the globe. Of these, 91 per cent are limited to a single country, which links their potential for extinction to national economic and social conditions. There are certainly other species, as the cases of ferns and fern allies in India, which have not yet been assessed.

The number of threatened vascular plant species recorded for each country. A high figure of threatened species for a particular country, like 29% in USA, generally indicates that inventories and assessments for that country have been particularly thorough, while a low figure for other countries like 7.7 % for India may reflect that similar efforts have not yet been undertaken there. The three countries, USA, Australia and South Africa, which were able to provide complete data sets, have higher percentages of threatened species in the national flora. Coverage of other regions is fragmentary and incomplete. It is necessary to make thorough assessment of flora in Asia, the Caribbean, South America, and the rest of Africa.

Out of 511 families of vascular plants currently recognized, 372 of these contain globally threatened and/or extinct species. Not surprisingly, the largest families also contain the largest numbers of threatened species. Excluding nineteen threatened monotypic families (only one species in the family, and thus 100 per cent threatened), there are 20 plant families with at least 50 per cent of their species threatened. Of these, eight are gymnosperm families (including cycads and conifers). The prominence of gymnosperms may be due to one or more factors:

1. They are a well known and relatively small group;
2. Many gymnosperm species are widely exploited both for timber and horticultural purposes;
3. Gymnosperms are an ancient group of species, and may not adapt easily to the rapidly changing environment around them.

In contrast, the ferns as a group appear to face relatively low levels of threat. This may be due in part to the efficiency with which fern spores are dispersed. At the same time, fern species have not been fully assessed, so their status as a group is not entirely clear. With this background it was planned to assess the Indian Pteridophytes, by selecting the Pteridophytes of the Western Ghats as the first step.

Methodology

Several methods, including the typical method adopted by IUCN, were tried to identify the rare and endangered Pteridophytes of the Western Ghats. As far as ferns are concerned, influences made by man are less except the destruction of forests. Very few ferns are over collected from the wild for economical use. With so small an area remaining under forest cover and the threat of further deforestation, about half of the pteridophyte species of southern India can be regarded as vulnerable, threatened or endangered. More or less all the ferns are growing in forest depth safely and they are not easy for access. Risk index rating of threatened ferns in Trinidad and Tobago has been done by Baksh-Comeau (1999) — by considering the number of Herbarium sheets available, date of first collection and last collection, total number of localities, total number of niches and distribution. The endangered to extinction species among the rare vascular plants in Poland are the taxa with limited area of habitats, taxa of an isolated sites by a significant disjunction from their compact ranges and endemics with limited number of specimens in their populations. Most of the existing methods, including the IUCN method, require a vast array of data particularly the field data that are rarely available for any group of plants in any country. Gathering of such data is also very difficult, because different kinds of data are to be collected by different experts. By considering these difficulties and lacunae in identifying the rare and endangered ferns, the key factors responsible for the rarity of the ferns were identified based on field studies and laboratory studies on south Indian ferns for the last twenty years.

For the successful establishment of a plant species in an ecosystem, the species should reproduce successfully through vegetative or sexual method by producing fertile and viable seeds/spores. It needs specific and suitable ecological niche to establish itself successfully. In the meantime they should have the capacity to colonise a particular ecological niche easily and in general they should have good genetical make up. As far as ferns are concerned the species with erect rhizome could not colonise a place easily when compared to the species with creeping rhizome. In general polyploid species are more tolerable than the diploid species. In evolution polyploidization is usually accomplished with the property of vegetative reproduction and thus polyploid species are usually with creeping rhizome while the diploid species are with erect rhizome. There are also some ferns with some intermediates like diploid with creeping rhizome. In addition, chlorophyllous spores are usually with short viability period when compared to other common brown coloured spores. Usually species with such chlorophyllous spores are rare. For example, all the members of the family Grammitidaceae and hymenophyllaceae with chlorophyllous spores are usually rare. Majority of ferns are terrestrials and they can grow easily by finding suitable places on the land surface. Some ferns are epiphytes or lithophytes, which need special hosts to grow. Due to the destruction of forests, several hosts of such epiphytes are vanished and without the host the epiphytes cannot grow. Thus usually diploid, epiphytic ferns with erect rhizome and chlorophyllous spores are very rare in contrast to the common polyploid, terrestrial ferns with creeping rhizome and achlorophyllous spores. There are also several intermediate combination of characters based on which the distribution of the species will vary.

Results and Discussion

With this context the key factors, particularly biological factors, such as rhizome type (erect or creeping), ploidy level (diploid or polyploid), nature of the spores (chlorophyllous or achlorophyllous) and ecological factor such as habitats, (epiphyte/lithophyte or terrestrial), responsible for the rarity of ferns and fern allies were identified and applied for the assessment of rare and endangered ferns and fern allies of the Western Ghats along with other criteria followed by Perring and Farrell. As far as ferns are concerned each and every species will score equally when we give the threat value for each criterion mentioned by Perring and Farrell like economical value, accessibility of the species etc. Because all the ferns are having at least some economical values and most of the ferns are growing in forest interior and it is very difficult for access. So it is very difficult to differentiate a rare and common fern when they score more or less same range of threat values. The application of the above biological criteria gave meaningful results and when they are applied separately they give more or less the same kind of results. So, for the present analysis only these criteria were used).

It is a very simple method based on only five criteria and the successfulness of this method has also been tested for ferns from other geographical regions (Himalayas) and it seems to be a natural, successful, easy method to test and locate a fern or fern ally in the red list category. The list of rare and endangered ferns of the Western Ghats, identified by following this method, is given here.

The validity of the present method has also been tested with the Himalayan ferns. In the first volume of *'An Illustrated Flora of Western Himalaya'* by Khullar (1994) totally 190 species of ferns have been described along with information on cytology, distribution and ecology. By applying the present method, all the 190 species have been categorized into different ranks. There are enumerated 46 rare and endangered ferns of the Western Ghats of South India, based on his own field experience for about 30 years. The species included in that list has also come under any one of the threat category of the present study and thus it is of more value. But the problem is those who want to identify a rare and endangered species for conservation purpose, he may not have such a kind of experience and he has to depend on either the ready made list or the scientific method to identify such species easily.

Few examples may be cited to test the validity of the present method. The diploid fern *Grammitis medialis* with erect rhizome, chlorophyllous spores and epiphyte/lithophyte habitat belonging to the first category has been recorded from only two localities from the Western Ghats. The diploid or tetraploid lithophytic fern *Hypodematium crenatum* with prostrate rhizome and achlorophyllous spores belonging to the first category has been recorded from only three distinct localitites from the Western Ghats.

In contrast, the tetraploid, terrestrial fern *Christella parasitica* with short or long-creeping rhizome, achlorophyllous spores belonging to the last category is the most common fern in South India. It is commonly growing throughout the Western Ghats. In the same way, the tetraploid, terrestrial bracken fern, *Pteridium aquilinum* with long creeping rhizome and

several forest areas and sanctuaries or biosphere reserves eg. KMTR, Nilgiri Biosphere reserves. For *ex situ* conservation special efforts are not given much to the cases of ferns when compared to the flowering plants. In majority of the gardens there are very few ferns which are grown mostly as ornamental ferns and not as a rare and endangered ferns. In India there are very few fernaries to conserve the rare and endangered ferns e.g. Kodaikanal Botanic Garden, Gurukula Botanic Garden, Nadugani Gene pool forests, and National Botanical Garden. The *ex situ* conservation of rare and endangered ferns may be strengthened by setting up more and more fernaries in different parts of the country particularly near by the sanctuary or biospheres. The *ex situ* conservation through *in vitro* tissue culture or spore culture has to be done at least in the species belonging to the first three ranks.

The purpose of the present chapter is to expose the rare and endangered ferns to the conservationists who are interested in conserving the rare and endangered ferns through *in vitro* tissue culture or spore culture. Usually they choose the species, for such conservation measures, without making serious efforts to identify the rare and endangered ferns. Some times they wrongly choose some of the common species even with the availability of the rare species. In India so far, ca out of 45 of species of ferns have been subjected to *in vitro* study with the aim of both academic studies and conservation studies. Based on the present study it has been observed that 45 of the selected species 31 species have been selected properly and 14 species have been selected wrongly. By following the present guidelines for the identification of rare and endangered ferns the wrong selection may be avoided and the correct species may be identified in future for conservation. Since Pteridologists have not been experienced in tissue culture and the biotechnologists are not aware of the taxonomy, ecology and distribution of the Pteridophytes, the present paper may serve as a link between these two groups of researchers.

2

Ferns

INTRODUCTION

A fern is any one of a group of about 20,000 species of plants classified in the phylum or division Pteridophyta, also known as Filicophyta. The group is also referred to as Polypodiophyta, or Polypodiopsida when treated as a subdivision of tracheophyta (vascular plants). The study of ferns and other pteridophytes is called pteridology, and one who studies ferns and other pteridophytes is called a pteridologist. The term "pteridophyte" has traditionally been used to describe all seedless vascular plants, making it synonymous with "ferns and fern allies". This can be confusing since members of the fern phylum Pteridophyta are also sometimes referred to as pteridophytes.

Ferns are vascular plants differing from the more primitive lycophytes by having true leaves (megaphylls), and they differ from seed plants (gymnosperms and angiosperms) in their mode of reproduction-lacking flowers and seeds. Like all other vascular plants, they have a life cycle referred to as alternation of generations, characterized by a diploid sporophytic and a haploid gametophytic phase. Unlike the gymnosperms and angiosperms, the ferns' gametophyte is a free-living organism. The life cycle of a typical fern is as follows:

— A sporophyte (diploid) phase produces haploid spores by meiosis;

— A spore grows by mitosis into a gametophyte, which typically consists of a photosynthetic prothallus;

— The gametophyte produces gametes (often both sperm and eggs on the same prothallus) by mitosis;

— A mobile, flagellate sperm fertilizes an egg that remains attached to the prothallus;

— The fertilized egg is now a diploid zygote and grows by mitosis into a sporophyte (the typical "fern" plant).

The stereotypic image of ferns growing in moist shady woodland nooks is far from being a complete picture of the habitats where ferns can be found growing. Fern species live in a wide variety of habitats, from remote mountain elevations, to dry desert rock faces, to bodies of water or in open fields. Ferns in general may be thought of as largely being specialists in marginal habitats, often succeeding in places where various environmental factors limit the success of flowering plants. Some ferns are among the world's most serious weed species, including the bracken fern growing in the British highlands, or the mosquito fern (*Azolla*) growing in tropical lakes, both species form large aggressively spreading colonies. There are four particular types of habitats that ferns are found in: moist, shady forests; crevices in rock faces, especially when sheltered from the full sun; acid wetlands including bogs and swamps; and tropical trees, where many species are epiphytes.

Many ferns depend on associations with mycorrhizal fungi. Many ferns only grow within specific pH ranges; for instance, the climbing fern (*Lygodium*) of eastern North America will only grow in moist, intensely acid soils, whilc thc bulblet bladder fern (*Cystopteris bulbifera*), with an overlapping range, is only ever found on limestone.

Fern Structure

— Ferns at the Royal Melbourne Botanical Gardens.

— Tree ferns, probably *Dicksonia antarctica*, growing in Nunniong, Australia.

Like the sporophytes of seed plants, those of ferns consist of:

- *Stems:* Most often an underground creeping rhizome, but sometimes an above-ground creeping stolon (e.g., Polypodiaceae), or an above-ground erect semi-woody trunk (e.g., Cyatheaceae) reaching up to 20 m in a few species (e.g., *Cyathea brownii* on Norfolk Island and *Cyathea medullaris* in New Zealand).
- *Leaf:* The green, photosynthetic part of the plant. In ferns, it is often referred to as a frond, but this is because of the historical division between people who study ferns and people who study seed plants, rather than because of differences in structure. New leaves typically expand by the unrolling of a tight spiral called a crozier or fiddlehead. This uncurling of the leaf is termed circinate vernation. Leaves are divided into three types:
  - *Trophophyll:* A leaf that does not produce spores, instead only producing sugars by photosynthesis. Analogous to the typical green leaves of seed plants.
  - *Sporophyll:* A leaf that produces spores. These leaves are analogous to the scales of pine cones or to stamens and pistil in gymnosperms and angiosperms, respectively. Unlike the seed plants, however, the sporophylls of ferns are typically not very specialized, looking similar to trophophylls and producing sugars by photosynthesis as the trophophylls do.
  - *Brophophyll:* A leaf that produces abnormally large amounts of spores. Their leaves are also larger than the other leaves but bear a resemblance to trophopylls.
- *Roots:* The underground non-photosynthetic structures that take up water and nutrients from soil. They are always fibrous and are structurally very similar to the roots of seed plants.

The gametophytes of ferns, however, are very different from those of seed plants. They typically consist of:

- *Prothallus:* A green, photosynthetic structure that is one cell thick, usually heart or kidney shaped, 3–10 mm long and 2-8 mm broad. The prothallus produces gametes by means of:

- *Antheridia:* Small spherical structures that produce flagellate sperm.
- *Archegonia:* A flask-shaped structure that produces a single egg at the bottom, reached by the sperm by swimming down the neck.

- *Rhizoids:* root-like structures (not true roots) that consist of single greatly-elongated cells, water and mineral salts are absorbed over the whole structure. Rhizoids anchor the prothallus to the soil.

One interesting difference between sporophytes and gametophytes might be summed up by the saying that "Nothing eats ferns, but everything eats gametophytes." This is an over-simplification, but it is true that gametophytes are often difficult to find in the field because they are far more likely to be food than are the sporophytes.

Evolution and Classification

Ferns first appear in the fossil record in the early-Carboniferous period. By the Triassic, the first evidence of ferns related to several modern families appeared. The "great fern radiation" occurred in the late-Cretaceous, when many modern families of ferns first appeared.

Ferns have traditionally been grouped in the Class Filices, but modern classifications assign them their own division in the plant kingdom, called Pteridophyta.

Traditionally, three discrete groups of plants have been considered ferns: two groups of eusporangiate ferns—families Ophioglossaceae (adders-tongues, moonworts, and grape-ferns) and Marattiaceae—and the leptosporangiate ferns. The Marattiaceae are a primitive group of tropical ferns with a large, fleshy rhizome, and are now thought to be a sibling taxon to the main group of ferns, the leptosporangiate ferns. Several other groups of plants were considered "fern allies": the clubmosses, spikemosses, and quillworts in the Lycopodiophyta, the whisk ferns in Psilotaceae, and the horsetails in the Equisetaceae. More recent genetic studies have shown that the Lycopodiophyta are only distantly related to any other vascular plants, having radiated evolutionarily at the base of the vascular plant clade, while both the whisk ferns and horsetails are as much "true" ferns as are the Ophioglossoids and Marattiaceae. In fact, the whisk ferns and Ophioglossoids are demonstrably a clade, and the horsetails and Marattiaceae are arguably another clade. Molecular data—which remain poorly constrained for many parts of the plants' phylogeny—have been supplemented by recent morphological observations supporting the inclusion of *Equisetaceae* within the ferns, notably relating to the construction of their sperm, and peculiarities of their roots.

One possible means of treating this situation is to consider only the leptosporangiate ferns as "true" ferns, while considering the other three groups as "fern allies". In practice, numerous classification schemes have been proposed for ferns and fern allies, and there has been little consensus among them. A new classification by Smith et al. (2006) is based on recent molecular systematic studies, in addition to morphological data. This classification divides ferns into four classes:

- Psilotopsida
- Equisetopsida
- Marattiopsida
- Polypodiopsida

The last group includes most plants familiarly known as ferns. Modern research supports older ideas based on morphology that the Osmundaceae diverged early in the evolutionary history of the leptosporangiate ferns; in certain ways this family is intermediate between the eusporangiate ferns and the leptosporangiate ferns.

The complete classification scheme proposed by Smith et al. (2006; alternative names in brackets):

- Class Psilotopsida
  - Order Ophioglossales
  - Family Ophioglossaceae (incl. Botrychiaceae, Helminthostachyaceae)

- Order Psilotales
- Family Psilotaceae (incl. Tmesipteridaceae)

• Class Equisetopsida [=Sphenopsida]

- Order Equisetales
- Family Equisetaceae

• Class Marattiopsida

- Order Marattiales
- Family Marattiaceae (incl. Angiopteridaceae, Christenseniaceae, Danaeaceae, Kaulfussiaceae)

• Class Pteridopsida [=Filicopsida, Polypodiopsida]

- Order Osmundales
- Family Osmundaceae
- Order Hymenophyllales
- Family Hymenophyllaceae (incl. Trichomanaceae)
- Order Gleicheniales
- Family Gleicheniaceae (incl. Dicranopteridaceae, Stromatopteridaceae)
- Family Dipteridaceae (incl. Cheiropleuriaceae)
- Family Matoniaceae
- Order Schizaeales
- Family Lygodiaceae
- Family Anemiaceae (incl. Mohriaceae)
- Family Schizaeaceae
- Order Salviniales
- Family Marsileaceae (incl. Pilulariaceae)
- Family Salviniaceae (incl. Azollaceae)
- Order Cyatheales
- Family Thyrsopteridaceae

- Family Loxomataceae
- Family Culcitaceae
- Family Plagiogyriaceae
- Family Cibotiaceae
- Family Cyatheaceae (incl. Alsophilaceae, Hymenophyllopsidaceae)
- Family Dicksoniaceae (incl. Lophosoriaceae)
- Family Metaxyaceae
- Order Polypodiales
- Family Lindsaeaceae (incl. Cystodiaceae, Lonchitidaceae)
- Family Saccolomataceae
- Family Dennstaedtiaceae (incl. Hypolepidaceae, Monachosoraceae, Pteridiaceae)
- Family Pteridaceae (incl. Acrostichaceae, Actiniopteridaceae, Adiantaceae, Anopteraceae, Antrophyaceae, Ceratopteridaceae, Cheilanthaceae, Cryptogrammaceae, Hemionitidaceae, Negripteridaceae, Parkeriaceae, Platyzomataceae, Sinopteridaceae, Taenitidaceae, Vittariaceae)
- Family Aspleniaceae
- Family Thelypteridaceae
- Family Woodsiaceae (incl. Athyriaceae, Cystopteridaceae)
- Family Blechnaceae (incl. Stenochlaenaceae)
- Family Onocleaceae
- Family Dryopteridaceae (incl. Aspidiaceae, Bolbitidaceae, Elaphoglossaceae, Hypodematiaceae, Peranemataceae)
- Family Oleandraceae
- Family Davalliaceae

– Family Polypodiaceae (incl. Drynariaceae, Grammitidaceae, Gymnogrammitidaceae, Loxogrammaceae, Platyceriaceae, Pleurisoriopsidaceae)

Economic Uses

Ferns are not as important economically as seed plants but have considerable importance. Some ferns are used for food, including the fiddleheads of bracken, *Pteridium aquilinum*, ostrich fern, *Matteuccia struthiopteris*, and cinnamon fern, *Osmunda cinnamomea]*. *Diplazium esculentum* is also used by some tropical peoples as food.

Ferns of the genus *Azolla* are very small, floating plants that do not look like ferns. Called mosquito fern, they are used as a biological fertilizer in the rice paddies of southeast Asia, taking advantage of their ability to fix nitrogen from the air into compounds that can then be used by other plants.

A great many ferns are grown in horticulture as landscape plants, for cut foliage and as houseplants, especially the Boston fern (*Nephrolepis exaltata*). The Bird's Nest Fern, *Asplenium nidus*, is also popular, and the staghorn ferns, genus *Platycerium*, have a considerable following.

Several ferns are noxious weeds or invasive species, including Japanese climbing fern (*Lygodium japonicum*), mosquito fern and sensitive fern (*Onoclea sensibilis*). Giant water fern (*Salvinia molesta*) is one of the world's worst aquatic weeds. The important fossil fuel coal consists of the remains of primitive plants, including ferns.

Ferns have been studied and found to be useful in the removal of heavy metals, especially arsenic, from the soil.

Other ferns with some economic significance include:

- *Dryopteris filix-mas* (male fern), used as a vermifuge, and formerly in the US Pharmacopeia; also, this fern accidentally sprouting in a bottle resulted in Nathaniel Bagshaw Ward's 1829 invention of the terrarium or Wardian case.

- *Rumohra adiantoides* (floral fern), extensively used in the florist trade.
- *Osmunda regalis* (royal fern) and *Osmunda cinnamomea* (cinnamon fern), the root fiber being used horticulturally; the fiddleheads of *O. cinnamomea* are also used as a cooked vegetable.
- *Matteuccia struthiopteris* (ostrich fern), the fiddleheads used as a cooked vegetable in North America.
- *Pteridium aquilinum* (bracken), the fiddleheads used as a cooked vegetable in Japan and are believed to be responsible for the high rate of stomach cancer in Japan. It is also one of the world's most important agricultural weeds, especially in the British highlands, and often poisons cattle and horses.
- *Diplazium esculentum* (vegetable fern), a source of food for some native societies.
- *Pteris vittata* (brake fern), used to absorb arsenic from the soil.
- *Polypodium glycyrrhiza* (licorice fern), roots chewed for their pleasant flavour.
- Tree ferns, used as building material in some tropical areas.
- *Cyathea cooperi* (Australian tree fern), an important invasive species in Hawaii.
- *Ceratopteris richardii*, a model plant for teaching and research, often called C-fern.

Cultural Connotations

Blätter des Manns Walfarn. by Alois Auer, Vienna: Imperial Printing Office, 1853.

In Slavic folklore, ferns are believed to bloom once a year, during the Ivan Kupala night. Although it's exceedingly difficult to find, anyone who takes a look of a fern flower will be happy and rich for the rest of his life. Similarly in Finland, the tradition holds that one who finds the seed of a fern in bloom on Midsummer night, will by the possession of it be able to travel under a glamour of invisibility and shall be guided to the locations where eternally blazing Will o' the wisps mark the spot of hidden treasure caches.

"Pteridomania"' is a term for the Victorian era craze of fern collecting and fern motifs in decorative art including pottery, glass, metals, textiles, wood, printed paper, and sculpture "appearing on everything from christening presents to gravestones and memorials." The fashion for growing ferns indoors led to the development of the Wardian case, a glazed cabinet that would exclude air pollutants and maintain the necessary humidity.

Fractal fern created using chaos game, through an Iterated function system (IFS).

The dried form of ferns was also used in other arts, being used as a stencil or directly inked for use in a design. The botanical work, *The Ferns of Great Britain and Ireland*, is a notable example of this type of nature printing. The process, patented by the artist and publisher Henry Bradbury, impressed a specimen on to a soft lead plate. The first publication to demonstrate this was Alois Auer's *The Discovery of the Nature Printing-Process*.

Medicinal Value

Ferns are sometimes used in medicine to treat cuts and clean them out. Ferns are also good bandages if you are stuck out in the wild. Rubbing a sword fern frond spore-side-down on a stinging nettle sting removes the stinging.

Misunderstood Names

Several non-fern plants are called "ferns" and are sometimes confused with true ferns. These include:

- *"Asparagus fern"* — This may apply to one of several species of the monocot genus *Asparagus*, which are flowering plants.
- *"Sweetfern"* — A flowering shrub of the genus *Comptonia*.
- *"Air fern"* — A group of animals called hydrozoan that are distantly related to jellyfish and corals. They are harvested,

dried, dyed green, and then sold as a "plant" that can "live on air". While it may look like a fern, it is merely the skeleton of this colonial animal.

- *"Fern bush"—Chamaebatiaria millefolium*—a rose family shrub with fern-like leaves.

In addition, the book *Where the Red Fern Grows* has elicited many questions about the mythical "red fern" named in the book. There is no such known plant, although there has been speculation that the oblique grape-fern, *Sceptridium dissectum*, could be referred to here, because it is known to appear on disturbed sites and its fronds may redden over the winter.

SPOROPHYTE

Young sporophytes of the common moss *Tortula muralis*. In mosses, the gametophyte is the dominant generation, while the sporophytes consist of sporangium-bearing stalks growing from the tips of the gametophytes.

All land plants, and some algae, have life cycles in which a haploid gametophyte generation alternates with a diploid sporophyte, the generation of a plant or alga that has a double set of chromosomes. A multicellular sporophyte generation or phase is present in the life cycle of all land plants and in some green algae. For common flowering plants (*Angiosperms*), the sporophyte generation comprises almost their whole life cycle (i.e. whole green plant, roots etc), except phases of small reproductive structures (pollen and ovule).

The sporophyte produces spores (hence the name), by meiosis. These meiospores develop into a gametophyte. Both the spores and the resulting gametophyte are haploid, meaning they only have one set of homologous chromosomes. The mature gametophyte produces male or female gametes (or both) by mitosis. The fusion of male and female gametes produces a diploid zygote which develops into a new sporophyte. This cycle is known as alternation of generations or alternation of phases.

In flowering plants, the sporophyte comprises the whole multicellular body except the pollen and embryo sac.

In the normal course of events, the zygote and sporophyte will have a full double set of chromosomes again. An exception is when a diploid and haploid gamete fuse, resulting in a triploid sporophyte, which will usually be sterile, as dividing three sets of chromosomes into two halves causes complications.

Bryophytes (mosses, liverworts and hornworts) have a dominant gametophyte stage on which the adult sporophyte is dependent on the gametophyte for nutrition. The embryo of the sporophyte develops from the zygote within the female sex organ or archegonium, and in its early development is therefore nurtured by the gametophyte. Because this embryo-nurturing feature of the life cycle is common to all land plants they are known collectively as the Embryophytes.

Most algae have dominant gametophyte generations, but in some species the gametophytes and sporophytes are morphologically similar (isomorphic). An independent sporophyte is the dominant form in all clubmosses, horsetails, ferns, gymnosperms, and angiosperms (flowering plants) that have survived to the present day. Early land plants had sporophytes that produced identical spores (isosporous or homosporous) but the ancestors of the gymnosperms evolved complex heterosporous life cycles in which the spores producing male and female gametophytes were of different sizes, the female megaspores tending to be larger, and fewer in number, than the male microspores.

During the Devonian period several plant groups independently evolved heterospory and subsequently the habit of endospory, in which single megaspores were retained within the sporangia of the parent sporophyte, instead of being freely liberated into the environment as in ancestral exosporous plants. These endosporic megaspores contained within them a miniature multicellular female gametophyte complete with female sex organs or archegonia containing oocytes which were fertilised by free-swimming sperm produced by windborne miniatuarised male gametophytes in the form of pre-pollen. The resulting zygote developed into the next sporophyte generation while still retained within the pre-ovule, the single large female meiospore or megaspore contained in the modified sporangium or nucellus of the parent sporophyte. The evolution of heterospory and endospory were among the earliest steps in the evolution of seeds of the kind produced by gymnosperms and angiosperms today.

EQUISETOPSIDA

Equisetopsida, or Sphenopsida, is a class of plants with a fossil record going back to the Devonian. Living species are commonly known as horsetails and typically grow in wet areas, with needle-like leaves radiating at regular intervals from a single vertical stem. Equisetopsida is placed in the botanical division of ferns (Pteridophyta) though sometimes regarded as a separate division Equisetophyta (also as Sphenophyta or Arthrophyta).

The Sphenophytes comprise photosynthesising, "segmented", hollow stems, sometimes filled with pith. At the junction between each segment is a whorl of leaves. In the only extant genus *Equisetum*, these are small leaves (microphylls) with a singular vascular trace. However, sphenophyte leaves probably arose by the reduction of a megaphyll, as evidenced by early fossil forms such as *Sphenophyllum*, in which the leaves are broad with branching veins. The plumbing of these leaves is interesting: the vascular traces trifurcate at the junctions, with one thread going to the microphyll, and the other two moving left and right to merge with the new branches of their neighbours. The vascular system itself curiously resembles that of the vascular plants' eustele, which evolved convergently. A primary xylem contains carinal canals; in the Calamitales, secondary xylem (but not secondary phloem) can be secreted as the cambium grows outwards, producing a woody stem, and allowing the plants to grow as high as 10m. The cortex itself contains valecular canals; due to the softer nature of the phloem, these are very rarely seen in fossil instances.

The plant does not bear a coherent root system but underground rhizomes, from which roots and aerial axes emerge. The plant contains an intercalary meristem: that is to say, each segment of the stem grows as the plant gets taller. This contrasts with the seed plants, which contain an apical meristem - i.e. new growth comes only from growing tips (and widening of stems). Growth was determinate - i.e. the plants' phenotype dictated a maximum height, which the plant would grow to then get no higher.

Sphenophytes bear cones (technically *strobili*, sing. *strobilus*) at the tips of some stems. These cones comprise spirally arranged sporophores, which bear spores in four clusters, and in extant sphenophytes cover the spores externally - like four sacs hanging from an umbrella, with its handle embedded in the central cone body. In extinct groups, further protection was afforded to the spores by the presence of whorls of bracts - big pointy microphylls protruding from the cone.

The spores themselves bear characteristic elaters, distinctive spring-like attachments which are hygroscopic: i.e. they change their configuration in the presence of water, helping the spores move and aiding their dispersal. Dispersal is aided in the first instance by laterally dehiscing sporangia, which pop open and scatter spores.

Vegetative stem: N = node, I = internode, B = branch in whorl, L = fused microphylls.

Cross-section through a strobilus; sporophores, with attached congregations of spores, can be discerned.

Strobilus of *E. telmateia*, terminal on an unbranched stem.

The extant horsetails are mostly homosporous, but this is conspicuously not the case in the past.

Fossil Record

The extant horsetails represent a tiny fraction of Sphenophyte diversity in the past. There were three orders of Equisetopsid; the Pseudoborniales, which first appeared in the late Devonian. Second, the Sphenophyllales which were a dominant member of the Carboniferous understory, and prospered until the mid and early Permian respectively. The Equisetales existed alongside the Sphenophyllales, but diversified as that group disappeared into extinction, gradually dwindling in diversity to today's single genus *Equisetum*.

The organisms first appear in the fossil record during the late Devonian a time when land plants were undergoing a rapid diversification, with roots, seeds and leaves having only just evolved. However, plants had already been on the land for almost a hundred million years, with the first evidence of land plants dating to 475 million years ago.

Systematics

The horsetails and their fossil relatives have long been recognized as quite distinct from other seedless vascular plants. In fact, the group is so unlike other living and fossil plants that its relationship to other plants has long been considered problematic.

Because of the unclear relationships of the group, the rank botanists assign to it varies from order to division. When recognized as a separate division, the literature uses many possible names, including Arthrophyta Sphenophyta , or Equisetophyta. Other authors have regarded the same group as a class, either within a division consisting of the vascular plants or, more recently, within an expanded fern group. When ranked as a class, the group has been termed the Equisetopsida or Sphenopsida.

MARATTIOPSIDA

Class Marattiopsida is a group of ferns containing a single order, Marattiales, and family, Marattiaceae. Class Marattiopsida diverged from other ferns very early in their evolutionary history and are quite different from many plants familiar to people in temperate zones. Many of them have massive, fleshy rootstocks and the largest known fronds of any fern. The Marattiaceae is one of two eusporangiate fern families, meaning that the sporangium is formed from a group of cells vs the leptosporangium in which there is a single initial cell. There are four extant genera (*Angiopteris*, *Christensenia*, *Danaea* and *Marattia*) and it has a long fossil history with many extinct taxa (*Psaronius*, *Asterotheca*, *Scolecopteris*, *Eoangiopteris*, *Qasimia*, *Marantoidea*, *Danaeites*, *Marattiopsis*, etc.)

In this group, such fronds are found in the genus *Angiopteris*, native to Australasia, Madagascar and Oceania. These fronds may be up to 9 meters long in the species *Angiopteris teysmanniana* of Java. In Jamaica the species *Angiopteris evecta* is widely naturalized and is registered as an invasive species. The plant was introduced by Captain Bligh from Tahiti as a staple food for slaves and cultivated in the Castleton Gardens in 1860. From there it was able to distribute itself throughout the eastern half of the island.

Another East-Asian genus is *Christensenia*, a peculiar fern with fronds resembling a horse chestnut leaf. That is why the species is called *Christensenia aesculifolia*, meaning Christensen's chestnut-leaf. Christensen was a famous Danish fern botanist.

The most widespread genus in Marattiaceae is the pantropical *Marattia*, usually occurring at higher elevations. These are also large ferns with globular rhizomes, but fronds can be up to 4 times pinnate. The sporangia are fused into bivalvate structures called a synangium. The New Zealand King Fern, *Marattia salicina*, known in Maori as Para also belongs here. Sometimes called the Potato Fern, this is a large Australasian fern with an edible fleshy rhizome that is used as a food source by some indigenous peoples.

The fourth genus *Danaea* is endemic to the Neotropics. They have bipinnate leaves with opposite pinnae, which are dimorphic, the fertile leaves much contracted, and covered below with sunken synangia.

Several other genera are known in the Marattiaceae, namely: *Archangiopteris*, *Macroglossum*, *Protangiopteris*, and *Protomarattia*. These are all synonyms of *Angiopteris*.

According to recent molecular studies it appears that these eusporangiate ferns are a sister group to the horsetails (Equisetaceae).

PTERIDOPSIDA

The Pteridopsida is a class of plants in the Division Pteridophyta that includes all the leptosporangiate ferns. The class is renamed Polypodiopsida. This recent reclassification of

Monilophyte (ferns) is based on multiple molecular studies published since 1994 that have clarified some of the confusion of the placement and relations among fern families. Polypodiopsida is one of four classes of Monilophytes (an Infradivision, this rank is not recognized by the *International Code of Botanical Nomenclature*), the other three being Marattiopsida, Equisetopsida, and Psilotopsida.

Discussion of Molecular Classification

There has been some challenge to the recent molecular studies, claiming that these provide a skewed view of the phylogenetic order because the studies don't take into account fossil representatives . However, the molecular studies have clarified relations among families that were thought to be non-monophyletic before the advent of molecular information, which were left in their non-monophyletic ranks because there was not enough information to do otherwise . The reclassification of ferns using multiple molecular studies, which have generally supported each other, is not any different from classifications of the past—it is the definition of the relations utilizing the all the information available. It does not discourage the further study and clarification of the groups, and does not mean that if further study proves the classification wrong, it will not be changed.

CYATHEALES

Scientific Classification

Kingdom	:	Plantae
Division	:	Pteridophyta
Class	:	Pteridopsida
Subclass	:	Cyatheatae
Order	:	Cyatheales

Families and Genera

Thyrsopteridaceae

Thyrsopteris

Loxomataceae

Loxoma

Loxsomopsis

Culcitaceae

Culcita

Plagiogyriaceae

Plagiogyria

Cibotiaceae

Cibotium

Cyatheaceae

Alsophila

Cyathea

Gymnosphaera

Hymenophyllopsis

Sphaeropteris

Dicksoniaceae

Calochlaena

Dicksonia

Lophosoria

Metaxyaceae

Metaxya

The order Cyatheales is a taxonomic division of the fern subclass, Cyatheatae, which includes the tree ferns.

In general, any fern that grows with a trunk elevating the fronds (leaves) above ground level can be called a tree fern. However, the plants formally known as tree ferns comprise a group of large ferns belonging to the families Dicksoniaceae and Cyatheaceae in the order Cyatheales.

Tree ferns are found growing in tropical and subtropical areas, including cool to temperate rainforests in Australia,

New Zealand, Malaysia, Lord Howe Island, and other island groups nearby; a few genera extend further, such as *Culcita* in southern Europe. Like all ferns, tree ferns reproduce by means of spores developed in sporangia on the undersides of the fronds.

Tasmanian tree fern (*Dicksonia antarctica*) in an English garden. The trunk is 60 cm (24 in) high.

The fronds of tree ferns are usually very large and multiple-pinnate but at least one type has entire (undivided) fronds. The fronds of tree ferns also exhibit circinate vernation, meaning the young fronds emerge in coils that uncurl as they grow.

Unlike flowering plants, tree ferns do not form new woody tissue in their trunk as they grow. Rather, the trunk is supported by a fibrous mass of roots that expands as the tree fern grows.

Some tree fern genera — for example *Dicksonia* and *Cibotium*, but not *Cyathea* — can be transplanted by severing the top portion from the rest of the trunk and replanting it. If the transplanted top part is kept moist it will regrow a new root system over the next year. The success rate of transplantation increases to about 80% if the roots are dug up intact. If the crown of the Tasmanian tree fern *Dicksonia Antarctica* (the most common species in gardens) is damaged, it will die because that is where all new growth occurs. But other clump-forming tree fern species, such as *D. squarrosa* and *D. youngiae*, can regenerate from basal offsets or from "pups" emerging along the surviving trunk length. Tree ferns often fall over in the wild, yet manage to re-root from this new prostrate position and begin new vertical growth.

Tree fern frond ("fiddlehead") by the Akatarawa River, New Zealand. These unopened fronds are edible but must be roasted first to remove shikimic acid.

It is not certain how many species of tree fern there are but it is likely to be around a thousand. More new species are discovered in New Guinea with each botanical survey. On the other hand, many species have become extinct in the last century as forest habitats have come under pressure from human intervention.

While many ferns are able to achieve a widespread distribution because of their spore reproduction, tree fern species tend to be very local. This makes their species much more susceptible to the effects of local deforestation. It is not known why species are not more widespread, especially considering that they have sufficient height to have a greater chance of getting spore into the wind stream.

Where feral pigs are a problem, such as in some Hawaiian forests, they often are able to knock over tree ferns and to root out the starchy pith, killing the plant

Outside of the Cyatheales a few ferns in other groups could be considered tree ferns, such as several ferns in the family Osmundaceae that can achieve short trunks under a metre tall and a few species in the genera *Blechnum, Leptopteris, Sadleria* and *Todea* could also be considered tree ferns in a liberal interpretation of the term.

The families that constitute Cyatheales have been relatively firmly established as a clade by DNA sequencing and morphological studies. The order Plagiogyriales, which contains the family Plagiogyriaceae, is most closely related to the Cyatheales, not to the Osmundales as had been previously supposed.

PTERIS

Scientific classification

Kingdom	:	Plantae
Division	:	Pteridophyta
Class	:	Pteridopsida
Order	:	Pteridales
Family	:	Pteridaceae
Genus	:	*Pteris*

Species

Pteris (brake) is a genus of about 280 species of ferns, native to tropical and subtropical regions of the world.

Many of them have linear frond segments, and some have sub-palmate division. Like other members of the Pteridales, the frond margin is reflexed over the marginal sori.

"Brake" has often been confused with "bracken". The relationship between the two words is uncertain. If they are related, then brake is a diminution of bracken.

Selected Species

teris aberrans Alderw.

Pteris abyssinica Hieron.

Pteris actiniopteroides Christ

Pteris adscensionis Sw.

Pteris albersii Hieron.

Pteris albertiae Arbelaez

Pteris altissima Poir.

Pteris amoena Bl.

Pteris angustata (Fée) C. Morton

Pteris angustipinna Tagawa

Pteris angustipinnula Ching & S.H.Wu

Pteris appendiculifera Alderw.

Pteris arborea L.

Pteris argyraea Moore

Pteris aspericaulis Wall. ex Hieron.

Pteris asperula J. Sm. ex Hieron.

Pteris atrovirens Willd.

Pteris auquieri Pichi-Serm.

Pteris austrosinica (Ching) Ching

Pteris bahamensis (Agardh) Fée

Pteris bakeri C. Chr.

Pteris baksaensis Ching

Pteris balansae Fourn.

Pteris bambusoides Gepp

Pteris barbigera Ching

Pteris barombiensis Hieron.

Pteris bavazzanoi Pichi-Serm.

Pteris beecheyana Ag.

Pteris bella Tagawa

Pteris berteroana Ag.

Pteris biaurita L.

Pteris biformis Splitg.

Pteris blumeana Agardh

Pteris boninensis H. Ohba

Pteris brassii C. Chr.

Pteris brevis Copel.

Pteris brooksiana Alderw.

Pteris buchananii Bak. ap. Sim.

Pteris buchtienii Rosenst.

Pteris burtonii Bak.

Pteris cadieri Christ

Pteris caesia Copel.

Pteris caiyangheensis L.L. Deng

Pteris calcarea Kurata

Pteris calocarpa (Copel.) M. Price

teris catoptera Kze.

Pteris chiapensis A. R. Sm.

Pteris chilensis Desv.

Pteris christensenii Kjellberg

Pteris chrysodioides (Fée) Hook.

Pteris ciliaris Eat.

Pteris clemensiae Copel.

Pteris comans Forst.

Pteris commutata Kuhn

Pteris concinna Hew.

Pteris confertinervia Ching

Pteris confusa T. G. Walker

Pteris congesta Prado

Pteris consanguinea Mett. ex Kuhn

Pteris coriacea Desv.

Pteris crassiuscula Ching & Wang

Pteris cretica L.

Pteris croesus Bory

Pteris cryptogrammoides Ching

Pteris cumingii Hieron.

Pteris dactylina Hook.

Pteris daguensis (Hieron.) Lellinger

Pteris dalhousiae Hook.

Pteris dataensis Copel.

Pteris dayakorum Bonap.

Pteris decrescens Christ

Pteris decurrens Presl

Pteris deflexa Link

Pteris deltea Ag.

Pteris deltodon Bak.

Pteris deltoidea Copel.

Pteris dentata Forsskal

Pteris denticulata Sw.

Pteris dispar Kze.

Pteris dissimilis (Fee) Chr.

Pteris dissitifolia Bak.

Pteris distans J. Sm.

Pteris droogmaniana L. Linden

Pteris edanyoi Copel.

Pteris ekmanii C. Chr.

Pteris elmeri Christ ex Copel.

·*Pteris elongatiloba* Bonap.

Pteris endoneura M. Price

Pteris ensiformis Burm.

Pteris esquirolii Christ

Pteris excelsa Gaud.

Pteris famatinensis Sota

Pteris fauriei Hieron.

Pteris finotii Christ

Pteris flava Goldm.

Pteris formosana Bak.

Pteris fraseri Mett. ex Kuhn

Pteris friesii Hieron.

Pteris gallinopes Ching

Pteris geminata Wall.

Pteris gigantea Willd.

Pteris glaucovirens Goldm.

Pteris goeldii Christ

Pteris gongalensis T. G. Walker

Pteris grandifolia L.

Pteris grevilleana Wall. ex Agardh

Pteris griffithii Hook.

Pteris griseoviridis C. Chr.

Pteris guangdongensis Ching

Pteris guizhouensis Ching

Pteris haenkeana Presl

Pteris hamulosa Christ

Pteris hartiana Jenm.

Pteris heteroclita Desv.

Pteris heteromorpha Fée

Pteris heterophlebia Kze

Pteris hillebrandii Copel

Pteris hirsutissima Ching

Pteris hirtula (C. Chr.) C. Morton

Pteris hispaniolica Maxon

Pteris holttumii C. Chr.

Pteris hondurensis Jenm.

Pteris hookeriana Ag.

Pteris hossei Hieron

Pteris hostmanniana Ettingsh

Pteris hui Ching

Pteris humbertii C. Chr.

Pteris hunanensis C.M.Zhang

Pteris inaequalis (Fée) Jenm

Pteris incompleta Cav.

Pteris inermis (Rosenstock) Sota

Pteris insignis Mett. ex Kuhn

Pteris intricata Wright

Pteris intromissa Christ

Pteris irregularis Kaulf

Pteris iuzonensis Hieron

Pteris izuensis Ching

Pteris johannis-winkleri C. Chr.

Pteris junghuhnii (Reinw.) Bak.

Pteris kawabatae Kurata

Pteris keysseri Rosenst

Pteris khasiana (Clarke) Hieron

Pteris kidoi Kurata

Pteris kinabaluensis C. Chr.

Pteris kingiana Endl.

Pteris kiuschiuensis Hieron.

Pteris laevis Mett.

Pteris lanceifolia Ag.

Pteris lastii C. Chr.

Pteris laurea Desv.

Pteris laurisilvicola Kurata

Pteris lechleri Mett.

Pteris lepidopoda M.Kato & K.U.Kramer

Pteris leptophylla Sw.

Pteris liboensis P.S.Wang

Pteris ligulata Gaud.

Pteris limae Brade

Pteris linearis Poir.

Pteris litoralis Rechinger

Pteris livida Mett.

Pteris loheri Copel.

Pteris longifolia L.

Pteris longipes D. Don

Pteris longipetiolulata Lellinger

Pteris longipinna Hayata

Pteris longipinnula Wall

Pteris luederwaldtii Rosenst

Pteris luschnathiana (Kl.) Bak.

Pteris luzonensis Hieron.

Pteris lidgatii (Bak.) Christ

Pteris macgregorii Copel

Pteris macilenta A. Rich.

Pteris maclurei Ching

Pteris maclurioides Ching

Pteris macracantha Copel

Pteris macrodon Bak.

Pteris macrophylla Copel.

Pteris macroptera Link

Pteris madagascarica Ag.

Pteris majestica Ching

Pteris malipoensis Ching

Pteris manniana Mett. ex Kuhn

Pteris melanocaulon Fée

Pteris melanorhachis Copel

Pteris menglaensis Ching

Pteris mertensioides Willd

Pteris mettenii Kuhn

Pteris micracantha Copel

Pteris microdictyon (Fée) Hook

Pteris microlepis Pichi-Serm

Pteris microptera Mett. ex Kuhn

Pteris mildbraedii Hieron

Pteris moluccana Bl.

Pteris monghaiensis Ching

Pteris montis-wilhelminae Alston

Pteris morii Masam

Pteris mucronulata Copel

Pteris multiaurita Ag.

Pteris multifida Poir.

Pteris muricata Hook

Pteris muricatopedata Arbelaez

Pteris muricella Fée

Pteris mutilata L.

Pteris natiensis Tagawa

Pteris navarrensis H. Christ

Pteris nipponica Shieh

Pteris novae-caledoniae Hook.

Pteris obtusiloba Ching & S.H.Wu

Pteris occidentalisinica Ching

Pteris olivacea Ching in Ching & S. H. Wu

Pteris opaca J. Sm.

Pteris oppositipinnata Fée

Pteris orientalis Alderw

Pteris orizabae M. Martens & Galeotti

Pteris oshimensis Hieron

Pteris otaria Bedd

Pteris pachysora (Copel.) M. Price

Pteris pacifica Hieron

Pteris paleacea Roxb

Pteris papuana Ces

Pteris parhamii Brownlie

Pteris paucinervata Fée

Pteris paucipinnata Alston

Pteris paulistana Rosenst

Pteris pearcei Bak

Pteris pedicellata Copel

Pteris pediformis M.Kato & K.U.Kramer

Pteris pellucida Presl

Pteris perrieriana C. Chr.

Pteris perrottetii Hieron

Pteris philippinensis Fée

Pteris phuluangensis Tag. & Iwatsuki

Pteris pilosiuscula Desv.

Pteris plumbea Christ

Pteris pluricaudata Copel.

Pteris podophylla Sw.

Pteris polita Link

Pteris polyphylla (Presl) Ettingsh.

Pteris porphyrophlebia C. Chr. & Ching

Pteris praetermissa T.G. Walker

Pteris preussii Hieron.

Pteris prolifera Hieron.

Pteris propinqua J. Agardh

Pteris pseudolonchitis Bory ex Willd.

Pteris pseudopellucida Ching

Pteris pteridioides (Hook.) Ballard

Pteris puberula Ching

Pteris pulchra Schlecht. & Cham.

Pteris pungens Willd.

Pteris purdoniana Maxon

Pteris purpureorachis Copel.

Pteris quadriaurita Retz.

Pteris quinquefoliata (Copel.) Ching

Pteris quinquepartita Copel.

Pteris radicans Christ

Pteris ramosii Copel.

Pteris rangiferina Presl ex Miq.

Pteris reducta Bak.

Pteris remotifolia Bak.

Pteris reptans T.G. Walker

Pteris rigidula Copel.

Pteris rosenstockii C. Chr.

Pteris roseo-lilacina Hieron.

Pteris ryukyuensis Tagawa

Pteris satsumana Kurata

Pteris saxatilis Carse

Pteris scabra Bory ex Willd.

Pteris scabripes Wall.

Pteris schlechteri Brause

Pteris schwackeana Christ

Pteris semiadnata Phil.

Pteris semipinnata L.

Pteris sericea (Fée) Christ

Pteris setigera (Hook. ex Beddome) Nair

Pteris setuloso-costulata Hayata

Pteris shimenensis C.M.Zhang

Pteris shimianensis H.S.Kung

Pteris silvatica Alderw.

Pteris similis Kuhn

Pteris simplex Holtt.

Pteris sintenensis (Masam.) C.M.Kuo

Pteris speciosa Mett. ex Kuhn

Pteris splendens Kaulf.

Pteris splendida Ching

Pteris squamaestipes C. Chr. & Tardieu

Pteris squamipes Copel.

Pteris stenophylla Wall. ex Hook. & Grev.

Pteris stridens Ag.

Pteris striphnophylla Mickel

Pteris subindivisa Clarke

Pteris subquinata (Wall. ex Bedd.) Agardh

Pteris subsimplex Ching

Pteris sumatrana Bak.

Pteris swartziana Ag.

Pteris taiwanensis Ching

Pteris talamauana Alderw.

Pteris tapeinidiifolia H.Itô

Pteris tarandus M.Kato & K.U.Kramer

Pteris tenuissima Ching

Pteris togoensis Hieron.

Pteris torricelliana Christ

Pteris trachyrachis C. Chr.

Pteris transparens Mett.

Pteris tremula R. Br.

Pteris treubii Alderw.

Pteris tricolor Linden

Pteris tripartita Sw.

Pteris tussaci (Fée) Hook.

Pteris umbrosa R. Br.

Pteris undulatipinna Ching

Pteris usambarensis Hier.

Pteris vaupelii Hieron.

Pteris venusta Kunze

Pteris verticillata (L.) Lellinger & Proctor

Pteris vieillardii Mett.

Pteris viridissima Ching

Pteris vitiensis Bak.

Pteris vittata L.

Pteris wallichiana Agardh

Pteris wangiana Ching

Pteris warburgii Christ

Pteris werneri (Rosenstock) Holtt.

Pteris whitfordii Copel.

Pteris woodwardioides Bory ex Willd.

Pteris wulaiensis C.M.Kuo

Pteris yakuinsularis Kurata

Pteris yamatensis (Tagawa) Tagawa

Pteris zahlbruckneriana Endl.

Pteris zippelii (Miq.) M.Kato & K.U.Kramer

Cultivation and Uses

Some of these ferns are popular in cultivation as houseplants. These smaller species are often called "table ferns".

Pteris vittata (commonly known as brake fern) was discovered to have the ability to "hyperaccumulate" (absorb large amounts of) arsenic from soil. The discovery was somewhat accidental, as the fern was growing at a central Florida site contaminated with large amounts of copper arsenate in the soil. Dr. Lena Q. Ma of the University of Florida later discovered that it had hyperaccumulated considerable amounts of arsenic from the soil. The discovery may lead to the use of *pteris vittata* as a potential bioremediation plant.

POLYPODIACEAE

Scientific classification

Kingdom	:	Plantae
Division	:	Pteridophyta
Class	:	Polypodiopsida
Order	:	Polypodiales
Family	:	Polypodiaceae

Genera

Over 60

Synonyms

Drynariaceae

Grammitidaceae

Gymnogrammitidaceae

Loxogrammaceae

Platyceriaceae

Pleurisoriopsidaceae

Polypodiaceae is a family of polypod ferns, which includes more than 60 genera divided into several tribes and containing around 1,000 species. Nearly all are epiphytes, but some are terrestrial. Drynariaceae, Grammitidaceae, Gymnogrammitidaceae, Loxogrammaceae, Platyceriaceae and Pleurisoriopsidaceae are nowadays usually included in the Polypodiaceae.

Their stems range from erect to long-creeping. The fronds are entire, pinnatifid, or variously forked or pinnate. The petioles lack stipules. The scaly rhizomes are generally creeping in nature.

Polypodiaceae species are found in wet climates, most commonly in rain forests. Notable examples include:

- *Pleopeltis polypodioides* (Resurrection fern)
- *Phlebodium aureum* (Golden serpent fern)

Pleopeltis polypodioides

Scientific classification

Kingdom	:	Plantae
Division	:	Pteridophyta
Class	:	Pteridopsida
Order	:	Polypodiales
Family	:	Polypodiaceae

Genus : *Pleopeltis*

Species : *P. polypodioides*

Polypodium polypodioides (Resurrection fern; syn. *Polypodium polypodioides*) is a species of creeping, coarse-textured fern native to the Americas and Africa.

The evergreen fronds of this fern are 25 cm high by 5 cm wide and monomorphic. Margins (edges) are mostly entire. The leathery, yellow-green pinnae (leaflets) are deeply pinnatifid, oblong to narrowly lanceolate, usually widest near middle, occasionally at or near base. It attaches to the limbs of its host plant with a branching, creeping, slender rhizome, which grows to 2 mm in diameter. The scales are lanceolate, with light brown base and margins, and having a dark central stripe.

Upper and Lower Sides of Fronds

The gametophytes (the haploid gamete producers) of this plant emerge from very small spores that float in the air and are deposited on moist tree branches. These spores are produced in sporangia that develop on the leaves of the fern's sporophyte. The fern can also reproduce by the division of its rhizomes.

On the underside of the blades, the sori (reproductive clusters) are round, discrete, and sunken. Their outline can be seen as raised dimples on the upper surface. They are typically near the outer edge, and occur on all but the lowest pinnae of fertile fronds. Indusium is absent. Sporangia are yellow to brown at maturity. Spores are produced from summer to fall.

Habitat and Water Absorption

This fern is an epiphyte, or air plant, which means it attaches itself to other plants and gets its nutrients from the air and from water and nutrients that collect on the outer surface of bark. The Resurrection fern lives on the branches of large trees such as cypresses and can often be seen carpeting the shady areas on limbs of large oak trees. However, it is known to grow on the surfaces of rocks and dead logs as well. It is often found in the company of other epiphytic plants such as Spanish moss.

The Resurrection fern gets its name because it can survive long periods of drought by curling up its fronds and appearing desiccated, grey-brown and dead. However, when just a little water is present, the fern will uncurl and reopen, appearing to "resurrect" and restoring itself to a vivid green color. It has been estimated that these plants could go 100 years without water and still revive after a single soaking.

When the fronds "dry", they curl with their bottom sides upwards. In this way, they can rehydrate the quickest when rain comes, as most of the water is absorbed on the underside of the leaf blades. Experiments have shown that they can lose almost all their free water and remain alive - up to 97%, though more typically they only lose around 76% in dry spells (Moran 2004). For comparison, most other plants would die after losing only 8-12%. This fern can lose almost all the water not hydrating the cells in its leaves and survive.

At least one study has shown association between *P. polypodioides* and moss indicating that this fern may rely on moss for some of its water needs.

AZOLLA

Scientific Classification

Kingdom : Plantae

Division : Pteridophyta

Class : Pteridopsida

Order : Salviniales

Family : Azollaceae

Wettst.

Genus : *Azolla*

Lam.

Species

Azolla caroliniana Willd.

Azolla filiculoides Lam.

Azolla japonica Franch. & Sav.

Azolla mexicana Presl

Azolla microphylla Kaulf.

Azolla nilotica Decne. ex Mett.

Azolla pinnata R.Br.

Azolla (mosquito fern, duckweed fern, fairy moss, water fern) is a genus of seven species of aquatic ferns, the only genus in the family Azollaceae. They are extremely reduced in form and specialized, looking nothing like conventional ferns but more resembling duckweed or some mosses.

Ecology

Azolla floats on the surface of water by means of numerous, small, closely-overlapping scale-like leaves, with their roots hanging in the water. They form a symbiotic relationship with the blue-green alga *Anabaena azollae,* which fixes atmospheric nitrogen, giving the plant access to the essential nutrient. This has led to the plant being dubbed a "super-plant", as it can readily colonise areas of freshwater, and grow at great speed - doubling its biomass every two to three days. The only known limiting factor on its growth is phosphorus, another essential mineral. An abundance of phosphorus, due for example to eutrophication or chemical runoff, often leads to *Azolla* blooms.

Their nitrogen-fixing capability of *Azolla* has led to it being widely used as a biofertiliser, especially in parts of southeast Asia. Indeed, the plant has been used to bolster agricultural productivity in China for over a thousand years. When rice paddies are flooded in the spring, they can be inoculated with *Azolla*, which then quickly multiplies to cover the water, suppressing weeds. The rotting plant material releases nitrogen to the rice plants, providing up to nine tonnes of protein per hectare per year.

Azolla are also serious weeds in many parts of the world, entirely covering some bodies of water. The myth that no mosquito can penetrate the coating of fern to lay its eggs in the water gives the plant its common name "mosquito fern."

Most of the species can produce large amounts of deoxyanthocyanins in response to various stresses ncluding bright sunlight and extremes of temperature, , causing the water surface to appear to be covered with an intensely red carpet. Herbivore feeding induces accumulation of deoxyanthocyanins and leads to a reduction in the proportion of polyunsaturated fatty acids in the fronds, thus lowering their palatability and nutritive value.

Azolla cannot survive winters with prolonged freezing, so is often grown as an ornamental plant at high latitudes where it cannot establish itself firmly enough to become a weed. It is not tolerant to salinity; normal plants can't survive in greater than 1-1.6%, and even conditioned organisms die in over 5.5% salinity.

Reproduction

Azolla reproduces sexually, and asexually by splitting.

Like all ferns, sexual reproduction leads to spore formation, but *Azolla* sets itself apart from other members of its group by producing two kinds. During the summer months, numerous spherical structures called sporocarps form on the undersides of the branches. The male sporocarp is greenish or reddish and looks like the egg mass of an insect or spider. It is two millimeters in diameter, and inside are numerous male sporangia. Male spores (microspores) are extremely small and are produced inside each microsporangium. Curiously, microspores tend to adhere in clumps called massulae.

Female sporocarps are much smaller, containing one sporangium and one functional spore. Since an individual female spore is considerably larger than a male spore, it is termed a megaspore.

Azolla has microscopic male and female gametophytes that develop inside the male and female spores. The female gametophyte protrudes from the megaspore and bears a small number of archegonia, each containing a single egg. The microspore forms a male gametophyte with a single antheridium which produces eight swimming sperm. The barbed glochidia on the male spore clusters are assumed to cause them to cling to the female megaspores, thus facilitating fertilization.

Use as Food

In addition to its traditional cultivation as a bio-fertilizer for wetland paddy, Azolla is finding increasing use for sustainable production of livestock feed Azolla is rich in proteins, essential amino acids, vitamins and minerals. Studies describe feeding azolla to dairy cattle, pigs, ducks, and chickens, with reported increases in milk production, weight of broiler chickens and egg production of layers, as compared to conventional feed. One FAO study describes how azolla integrates into a tropical biomass agricultural system, reducing the need for inputs.

Study of Arctic climatology reported that azolla may have had a significant role in reversing a greenhouse effect that occurred 55 million years ago that caused the region around the north pole to turn into a hot tropical environment. This research conducted by the Institute of Environmental Biology at Utrecht University claims that large dense patches of *Azolla* growing around freshwater lakes formed by the climate change eventually consumed enough carbon dioxide for the greenhouse effect to reverse.

3

Succulent Ferns

BOTRYCHIUM SPECIES

- *Botyrichium*, from the Greek botrus (*botrys*), "grape"
- Grape Fern, from the prominent clusters of round spore cases which resemble miniature clusters of grapes.
- Moonwort, from the "half moon" leaflets and the Anglo-Saxon *wort*, "plant"

Taxonomy

- Kingdom *Plantae*, the Plants
- Division *Polypodiophyta*, the True Ferns
- Class *Filicopsida*
- Order *Ophioglossales*
- Family *Ophioglossaceae*, the Adder's Tongue or Succulent Ferns
- Genus *Botrychium*, the Grape Ferns
- North Country Grape Ferns:
  - *acuminatum*, Tailed Grape Fern
  - *dissectum*, Cutleaf Grape Fern
  - *lanceolatum*, Lance Leaf Grape Fern (TH-MN)

- *maticariifolium*, Matricary Grapefern
- *multifidum*, Leathery Grapefern
- *simplex*, Small Grape Fern (SSC-MN)
- *virginianum*, Rattlesnake Fern
- North Country Moonworts
- *campestre*, Iowa Moonwort (SSC-MN)
- *lunaria*, Common Moonwort (TH-MN)
- *minganense*, Mingan Moonwort (SSC-MN)
- *pallidum*, Pale Moonwort (EN-MN)
- *pseudopinnatum*, False Daisyleaf Moonwort
- 50-60 species worldwide; 30 in North America.

Description

- A highly variable genus of small, rare (or, at least, much overlooked), un-fernlike ferns.
- *Sterile Frond* (trophophore) a single leaf, ascending to perpendicular to stem, with or without leaf stalk:
  - Blades linear, oblong, or deltate; simple to highly dissected; 1½″-10" × ½″-14"
  - Pinnae (leaflets) spreading to ascending, highly variable (fan-shaped to lanceolate to linear); edges smooth to toothed. Reduced to segments in many species.
- Fertile Frond (sporophore) a stalk terminating in a cluster of tiny ball-like spore cases (sporangia) that resemble a bunch of grapes (hence the generic common name of "Grape Fern").
- *Stem single*, upright, succulent, fragile, sometimes hollow. Several layers of leaf primordia at base of stem and just below the ground, from which will develop leaves of future seasons.

- *Roots thick*, spongy, occasionally branching laterally, yellowish to black, 0.5mm-2mm in diameter, smooth or with corky ridges, not proliferous.

Identification

- Identifiable as *Botrychium* by the single succulent stem bearing a single leaf and a single fertile stalk, topped with clustered spore cases.
- Field Marks:
  - succulent, fragile, sometimes hollow stem
  - single leaf, lobed or compound
  - fertile stalk, usually branched, and topped with clusters of spore cases
  - diminutive size (usually).

Botrychium ID for the North Country

The Grape Ferns tend to be small, rare, and easily overlooked. There is one exceptions in our area. The Rattlesnake Fern, *Botrychium virginianum* is much larger (12"-18" tall) and easily the most commonly seen Grape Fern in the North Country.

- Recognized as a Grape Fern by its large, single leaf and, in early summer, the single fertile stem arching above it.
- Distinguished from other northern Grape Ferns by its much larger size and the lacy, thin textured, non-leathery leaf.

If you have a small Grape Fern, look at the single leaf.

- If the leaf is broadly triangular, you have *Botrychium multifidum*, Leathery Grapefern. The leaf will also be relatively large (for a Grape Fern), with dense overlapping leaflets and long, prominent leaf stalk. The Dissected Grape Fern, *Botrychium dissectum*, a more southerly species, has been collected once in St. Louis County, and is differentiated by its highly dissected leaflets.
- If the leaf is twice cut into narrow and pointed segments (lance-shaped or lanceolate) you have *Botrychium*

lanceolatum, Lance Leaf Grape Fern, a Threatened Species in Minnesota. The leaf will be small (only about 1" long) and high on stalk, resembling a daisy leaf in shape. The fertile frond will be branched into several equally long branches.

- If the leaf is twice cut into broader lobes with more rounded tips, you have *Botrychium maticariifolium,* Matricary Grapefern. The leaf will be small (only about 1" long), oblong, and pale green; often clasping spore-bearing stalk. The fertile frond will be branched into several branches of unequal lengths. A close relative of the Matricary Grape Fern, *Botrychium acuminatum,* Tailed Grape Fern, is found in the Lake Superior region in North Ontario and the UP. It has been collected once in our area, in Cook County.
- If the leaf is simply compound leaf (undivided lobes), you almost certainly have the highly variable *Botrychium simplex,* Small Grape Fern, a Species of Special Concern in Minnesota. Further solidify the identification by the combination of very small size (leaf only about 1½" long, plant less than 5½" tall), unbranched fertile frond, and clasping, simply compound leaf.
- If the leaf is narrow and upright, made up of pairs of fan shaped leaflets, you have a Moonwort.
- If the leaflets are broad and often overlapping, then you have *Botrychium lunaria,* Common Moonwort, a Threatened Species in Minnesota.
- If the leaflets are narrow and widely spaced, then you have *Botrychium minganense,* Mingan Moonwort, a Species of Special Concern in Minnesota.
- If the leaflets are often folded longitudinally and pale green to whitish in color, then you have *Botrychium pallidum,* Pale Moonwort, an Endangered Species in Minnesota.
- Two other species of Moonwort have been collected but once in NE Minnesota, Iowa Moonwort, *Botrychium campestre,* and False Daisyleaf Moonwort, *Botrychium pseudopinnatum.*

Distribution

- The greatest diversity in *Botrychium* is at high latitudes and high elevations, mostly in disturbed meadows and woods.
- Our North Country species are widely distributed across northern North America.
- Many species are circumpolar, occurring in Europe and Asia as well.
- Our 12 North Country species have all been recorded in Northeastern Minnesota, but only the Rattlesnake Fern (*Botrychium virginianum*) is at all common, the others ranging from rare to extremely rare.

Reproduction

- By means of microscopic spores.
  - Spore germinates, developing into a tiny underground structure (gametophyte) that produces the gametes (egg and sperm).
  - Mature sperm is released from one part of the gametophyte and swims via a thin film of water to the egg.

    Fertilized zygote then develops roots, stem, and the above-ground structure recognizable as a fern (sporophyte).
  - Sporophyte produces the spores by the thousands in round sacs (sporangia) borne in clusters at the top of the fertile stalk.
- Growth rate is slow; typically only a single leaf is produced each year. Primordia for several years are contained within the bud, but only one matures each season.
- Several species undergo periods of dormancy, where the plant will not appear for one to several years and then re-emerge in the exact same location.

Osmunda Cinnamomea

- *Osmunda*, from the Saxon god Osmunder the Waterman, who hid his family from danger in a clump of these ferns.
- *cinnamonea*, from the Latin "cinnamon"
- Common name from the wooly, cinnamon colored spores borne on the fertile frond.
- Other common names include Buckhorn, *Osmonde Cannelle* (Qué), *yamadori-zenmai* (Jpn)

Taxonomy

- Kingdom *Plantae*, the Plants
- Division *Polypodiophyta*, the True Ferns
- Class *Filicopsida*
- Order *Polypodiales*
- Family *Osmundaceae*, the Flowering Ferns (an oxymoron, of course)
- Genus *Osmunda*, the Flowering Ferns
- Taxonomic Serial Number: 17219
- *Osmunda* ferns form spores on a modified frond. For this species, a spore-bearing frond grows from the rhizome separate from sterile fronds.

Description

- A tall, deciduous, perennial fern of moist woods, 2'-5' tall.
- Fronds annual, erect, dimorphic.
- Sterile Fronds light green, ovate to lanceolate, up to 5' tall and 6"-12" wide, forming symmetric clump, turning yellow then bronze in fall. Pinnae (leaflets) broadly oblong with persistent tuft of hairs on upper surface at base; deeply lobed, almost to rachis but lacking true pinnules.
- Fertile Fronds bright green, turning to rich cinnamon brown; shorter and narrower than sterile leaves with much smaller, nonphotosynthetic pinnae, withering after sporulation.

- Stem round and slightly grooved; at first covered with cinnamon-colored hairs, later smooth and green.
- Rootstalk stout, woody, and creeping to suberect:
  - Roots fibrous, matted.
  - Fiddleheads large, showy with silver-white hairs which turn rusty as fronds unfurl.

Identification

- Unmistakeable when fertile cinnamon fronds are present.
- Distinguished from Ostrich Fern () by small tufts of rusty woolly hairs at the base of each pinna where the pinna meets the frond.
- Field Marks:
  - distinctive, cinnamon-colored fertile, spore-bearing frond
  - twice-cut sterile fronds
  - utter absence of sori on green leaves

Distribution

- Ontario to Newfoundland, south to Texas, Oaxaca, Veracruz, the Gulf Coast, and Florida; Cuba, Jamaica and Puerto Rico.
- Also Chiapas and Tabasco, south to Ecuador, Paraguay, and Brazil. The Russian Far East, Korea, China, Japan,Taiwan, northern Burma, NE Thailand, and Vietnam.

Habitat

- Bogs, peatlands, thickets, wet woods, swamps, ditches, and streambanks.
- On poorly drained, acid soils with high organic content.

Fire

- Fronds are killed by fire but the fern resprouts from rhizome. Has good fire tolerance, often showing vigorous rhizome growth after fire.

- Spores germinate on mineral soil, so it probably colonizes after fire.
- Where fire is of intensity or duration that it completely consumes the organic substrate, the fern will not survive.

Associates

- Sphagnum Mosses (*Sphagnum* sp.)

Uses

- Fiddleheads up to 8" are sometimes collected in the spring, steamed or boiled, and eaten.

Reproduction

- Reproduces by spores and vegetatively by rhizomes.
- Spores have very short viability after release and either fail to germinate or germinate poorly after just a few weeks.
- Both sterile and fertile fronds expand during a short period in early spring. Leaf expansion is complete within about a month. Fertile fronds begin to wither in early summer, after sporulation is completed. Sterile pinnae begin to wither at the end of summer, and the leaf stalk somewhat later, until the entire above ground plant is dry.

Propagation

- By rhizome division in fall or spring; however this species tends to take a long time to recover after division.
- By spores sown as soon as they ripen in mid to late summer.

Cultivation

- Widely cultivated as an ornamental. Hardy to USDA Zone 3 (average minimum annual temperature -40°F)
- Cultural Requirements
  - Full sun if constantly wet; otherwise part shade in moist soil
  - Organic, acidic soil (pH 4.5-7.0)

– Leave dead fronds over winter to protect crown.

- 3'-5' High x 2'-3' Wide
- Good for naturalizing in wet woodlands and as background for smaller, more colorful flowering plants.
- Available by mail order from specialty suppliers or at local nurseries.

ASPLENIUM TRICHOMANES

- *Asplenium,* from the Latin *splen,* "spleen".
- *trichomanes,* the Greek name (tricomanes) for this fern, from qric (*thrix*), "hair of the head".
- Common Name, an anglicized version of the Greek genus & species names.
- Other common names include *Doradille Chevelue* (Qué), *Svartbräken, Bergspring, Stenbräken, Vanlig Svartbräken* (Swe), *Svartburkne* (Nor), *Rundfinnet Radeløv* (Dan), *Tummaraunioinen* (Fin), *Brauner Streifenfarn* (Ger), *Aranyos Fodorka* (Hun).

Taxonomy

- Kingdom *Plantae,* the Plants
- Division *Polypodiophyta,* the True Ferns
- Class *Filicopsida*
- Order *Polypodiales*
- Family *Aspleniaceae,* the Spleenworts
- Genus *Asplenium,* the Spleenworts
- Taxonomic Serial Number: 17364
- Also known as *Asplenium melanocaulon*
- In Europe and North America, occurs as diploid and tetraploid cytotypes, generally treated as subspecies.
- *Asplenium trichomanes* ssp. *trichomanes,* the diploid, is found on noncalcareous rocks. This is the subspecies of our area.

- *Asplenium trichomanes* ssp. *quadrivalens*, the tetraploid, grows on calcareous substrates in northeastern North America.
- Triploid hybrids are known between the diploids and tetraploids.

Description

- **Fronds** evergreen, ¾" wide x 3"-6" long, growing in neat rosettes. Fertile and sterile fronds alike.
  - *Petiole* reddish brown or blackish brown throughout, lustrous, ½"-1½" long, 1/6-1/4 length of blade.
  - *Blade* linear, once-cut, ¼"-¾"×1¼"-8½", thin, smooth or sparsely haired; base gradually tapered; tip narrowly acute.
  - *Rachis* (axis) reddish brown throughout, lustrous, and smooth or nearly so.
  - *Pinnae* in 15-35 pairs, oblong to oval; edges shallowly toothed to more or less smooth; tip obtuse.
  - *Pinnules* absent
  - *Sori* 2-4 pairs per pinna
- Rootstalk short-creeping, often branched; scales black throughout or with brown borders, lanceolate.
  - Roots not proliferous.

Identification

- Unlike anything else in the North Country.
- Distinguished from other small, rock-loving ferns by its once-cut fronds (all other small rock ferns, excepting the Common Polypody, *Polypodium virginianum*, have more greatly dissected fronds.) Distinguished from the Polypody by its prominent, dark frond stem and the narrow fronds of nearly constant width.
- *Field Marks*
  - once-cut fronds
  - prominent, dark frond stem

- fondness for rocky places
- small size

Distribution

- Alaskan panhandle south to Oregon; the Black Hills south to Arizona, New Mexiço, and the Big Bend; North Ontario to Newfoundland, south to northeastern Minnesota, Wisconsin, Illinois, Missouri, Arkansas, and the southern Appalachians.
- Also Chihuahua, Europe, Asia, Africa, and Australia.

Habitat

- Acidic rocks such as sandstone, basalt, and granite; very rarely on calcareous rocks in our area.

Reproduction

- By spore and vegetatively by rhizome

Propagation

- By rhizome division

Cultivation

- Hardy to USDA Zone 3 (average minimum annual temperature -40°F)
- Good for rock gardens, rock walls
- Available by mail order from specialty suppliers
- Likes well-drained, moist, limy loam, some sun.

Diphasiastrum Species

- *Diphasiastrum*, from the generic name Diphasium, and *astrum*, "inferiority or partial resemblance", hence, "false Diphasium"
- Ground Cedar, from the resemblance of its scale-like leaves to those of the Cedars
- Other common names include Ground Pine, Clubmoss

Taxonomy

- Kingdom *Plantae,* the Plants
- Division *Lycopodiophyta,* the Clubmosses
- Class *Lycopodiopsida,* the Clubmosses
- Order *Lycopodiales,* the Clubmosses
- Family *Lycopodiaceae,* the Clubmosses
- Genus *Diphasiastrum,* the Ground Cedars
- Also known as *Lycopodium, Diphasium*
- North Country Species:
  - *complanatum,* Ground Cedar
  - *digitatum,* Fan Clubmoss
  - *tristachyum,* Blue Ground Cedar
  - Five species and six fertile hybrids occur in North America.

Description

- A creeping, evergreen, rhizomatous clubmoss, with occasional erect branches topped with slim cones.
- Vertical stems multi-branched with scale-like leaves
- Horizontal stems at or below surface of ground.
- Cones cylindrical, ½"-2½", in upright candelabra-like clusters

Identification

- Club Moss Identification for Amateurs

Our North Country clubmosses are divided into four genera, based upon their leaves and cones.

  - Clubmosses with flat, scale-like leaves belong to this genus, *Diphasiastrum*, the so-called Ground Cedars, represented by three species in the North Country. All other clubmosses in our area have pointed leaves and a bristly appearance.

- Clubmosses without cones belong to the genus *Huperzia*, represented by four species in the North Country, only two at all common.
- A clubmoss with a cone which is merely a somewhat broader extension of the shoot belongs to the genus *Lycopodiella*, represented by a single species in the North Country.
- Clubmosses with distinct cones and bristly leaves belong to the genus *Lycopodium*, the so-called Ground Pines, represented by six species in the North Country.

If you have a *Diphasiastrum*, look closely at the ultimate branchlets. If they are cord-like, and nearly square in cross section, then you have Blue Ground Cedar (*Diphasiastrum tristachyum*). Confirm by:

- bluish colour (hence the common name)
- leaves on the underside of the branchlet more or less the same size as those on the top and sides

If the ultimate branchlets are narrowly blade-like and flat in cross section, usually green; underside leaves much smaller than lateral and upperside leaves, look at the overall pattern of the branchlets.

- If the branchlets are irregular and almost shaggy in form, you have the common Ground Cedar (*Diphasiastrum complanatum*). Confirm by:
  - conspicuous annual bud constrictions
  - cone stalks, if present, regularly forked
  - cones mostly ½"-1", lacking sterile tips
- If the branchlets are very regularly fan-shaped, you have Fan Clubmoss (*Diphasiastrum digitatum*). Confirm by:
  - absence of conspicuous annual bud constrictions
  - cone stalks mostly branching abruptly at base
  - cones mostly ¾"-1½", many with sterile tips

Reproduction

- Clonal, reproducing primarily by sprouting from rhizomes.

Propagation

- Very difficult.

Cultivation

- Clubmosses can make attractive ground covers, but they do not transplant well.

Woodsia Species

- *Woodsia,* for English botanist Joseph Woods (1776-1864).
- Common name from the preferred habitat of these species.
- Other common names include Woodsia, *Hällebräknar* (Swe), *Kiviyrtit* (Fin), *Wimperfarn* (Ger)

Taxonomy

- Kingdom *Plantae,* the Plants
- Division *Polypodiophyta,* the True Ferns
- Class *Filicopsida*
- Order *Polypodiales*
- Family *Dryopteridaceae,* the Wood Ferns
- Genus *Woodsia,* the Cliff Ferns
- Taxonomic Serial Number: 17736
- North Country *Woodsia* species:
  - *alpina,* Alpine Woodsia (Species of Special Concern-MN)
  - *glabella,* Smooth Woodsia (Threatened Species-MN)
  - *ilvensis,* Rusty Woodsia
  - *oregana,* Oregon Woodsia
  - *scopulina,* Rocky Mountain Woodsia (Threatened Species-MN)

- About 30 species worldwide, mostly north temperate regions and higher elevations in the tropics.
- Genus *Cystopteris* probably the closest relative.
- North American *Woodsia* fall into two natural groups that might be recognized as subgenera.
- *Circumboreal Woodsia*, (*Woodsia ilvensis* , *Woodsia glabella* and *Woodsia alpina*), show clear affinities to Eurasian species and are marked by:
  - articulate petioles
  - smooth or crinkled edge leaflets (pinnules)
  - uniformly colored stem scales
  - indusial segments uniseriate throughout
  - chromosome base numbers of 39-41
- *Endemic Woodsia*, (*Woodsia oregana* and *Woodsia scopulina*) are found only in our hemishpere, and are marked by:
  - petioles that are not articulate
  - toothed edge leaflets (pinnules)
  - often bicolored stem scales
  - indusial segments multiseriate at the base
  - chromosome base number of 38
- *Hybridization* is common within these two groups, but intergroup hybrids are relatively rare.

Description

- A group of small ferns of rocky places.
- Fronds monomorphic, either deciduous or evergreen, erect to ascending from the ground-level stems, unbranched, but rarely horizontal; arising close together amid a cluster of persistent petiole bases; usually less than 6" high.
  - Petiole 1/5-3/4 length of blade, base not conspicuously swollen; smooth or more often hairy; hairs more than the diameter of petiole.

- Blades 1¼"-6" (rarely to 10") long, linear or lanceolate when mature, with inconspicuous veins; most species hairy.

- *Rookstalks* compact to creeping, ascending or erect (rarely horizontal), stolons absent.
- *Roots black*
- *Indusium* of narrow, hair-like segments encircling sorus, one row of cells many times longer than wide, and longer than the sporangia; persistent but often obscure in mature sori.
- *Spores brownish*

Identification

- Identifiable as *Woodsia* by:
  - articulate bases to the petioles and the accumulation of petiole bases that have broken off below the articulation.
  - relatively small size for our area
  - affinity for rocky habitats
- Distinguished from other small ferns of rocky places by:
  - twice-cut fronds
- Field Marks
  - hairs and scales on fronds and leafstalks
  - presence or absence of stem segmentation or articulation
  - colour of mature leaf stalk
- Woodsia Identification for Amateurs

Look for the scattered hairs and scales characteristic of this genus. If you don't find any, look for dark-colored leaf stalks. If all you can find are smooth greenish stalks with smooth blades, then you have *Woodsia glabella* or Smooth Woodsia. This will be a very small fern, with fronds less than ½" wide and fan-shaped lower leaflets. This fern is classified as Threatened in Minnesota and has been found in Cook and Lake Counties, but not St. Louis.

If you have a more typical *Woodsia,* check the leaf stalk to see if it is segmented, with a node clearly visible well above the base. In our area, we have two *Woodsia* with a segmented or articulated leaf stalk, the robust *Woodsia ilvensis* or Rusty Woodsia, and the more delicate *Woodsia alpina,* or Mountain Woodsia.

- Rusty Woodsia is our most common Woodsia and has abundant hairs and scales. Its largest leaflets (pinnae) have 4-9 pairs of secondary leaflets (pinnules).
- Alpine Woodsia is much less common, classified as a "Species of Special Concern" in Minnesota and found in Cook and Lake Counties, but not St. Louis. It has only scattered hairs and scales, and its largest pinnae have but 1-3 pairs of pinnules.

If, on the other hand, your more typical Woodsia has an unsegmented leaf stalk, it is one of our two remaining species, *Woodsia scopulina* (Rocky Mountain Woodsia) or *Woodsia oregana* (Oregon Woodsia).

- Rocky Mountain Woodsia is classified as a Threatened species in Minnesota and in our area is known only from Cook County in the far northeastern corner of the state. It is best identified by its hairs which are concentrated along the midrib on both surfaces, and by its mature leafstalks, which are typically a reddish brown to dark purple in color and relatively brittle, shattering easily.
- Despite its western name, Oregon Woodsia occurs throughout our area. Unlike its less common cousin (with another western name), it lacks a concentration of hairs along the midrib. Its mature leafstalks are often light brown to straw-colored and though occasionally reddish brown to dark purple they remain much more pliable than the leafstalks of the Rocky Mountain Woodsia, and resistant to shattering.

In the event, frustrating though it may be, that your specimen seems to fall somewhere between these species characteristics, you may have stumbled across a Woodsia hybrid.

This genus of fern is notorious for hybridization and hybrids of Rocky Mountain Woodsia and Oregon Woodsia (*Woodsia* x *maxonii*), Oregon Woodsia and Rusty Woodsia (*Woodsia* x *abbeae*), and Rusty Woodsia and Alpine Woodisa (*Woodsia* x *gracilis*) have all been found in northeastern Minnesota. Identification of the hybrids is beyond the scope of this effort, other than to note that the hybrids typically have characteristics midway between those of the two parent species.

Distribution

- Circumpolar. Arctic, or low arctic, or alpine.

Habitat

- Usually growing on rock

Propagation

- By rhizome division

Cultivation

- Hardy to USDA Zone 3 (average minimum annual temperature -40°F)
- Rarely available commercially.

THELYPTERIS PALUSTRIS

- *Thelypteris,* from the Greek, qelus (*thelus*), "female", and *pteris* (*pteris*), "fern"
- *palustris,* from the Latin, "fenny, marshy, swampy"
- Common Name, from its preferred habitat
- Other common names include Eastern Marsh Fern, Meadow Fern, *thélyptère des marais* (Qué), *Kärrbräken* (Swe), *Myrtelg* (Nor), *Kær-Dunbregne* (Dan), *Nevaimarre* (Fin), *Sumpffarn* (Ger), *Raineach Lèana* (Gaelic), *Harilik Soosõnajalg, Sõnajalg, Soosõnajalg* (Estonia), *Tõzegpáfrány* (Hun)

Taxonomy

- Kingdom *Plantae,* the Plants
- Division *Polypodiophyta,* the True Ferns

- Class *Filicopsida*
- Order *Polypodiales*
- Family *Thelypteridaceae,* the Marsh Ferns
- Genus *Thelypteris,* the Lady Ferns
- Also known as *Acrostichum thelypteris, Dryopteris thelypteris, Lastrea thelypteris, Thelypteris confluens* var. *pubescens* , *Thelypteris thelypteroides*
- Varieties
- *Thelypteris palustris* var. *palustris* occurs in Eurasia.
- *Thelypteris palustris* var. *pubescens* occurs in Eastern North America.

Description

- A common wetland fern; each frond arising individually from a creeping rhizome without forming clumps.
- Fronds monomorphic or slightly dimorphic, deciduous; fertile leaves more erect, narrower, with leaf edges slightly rolled over spore-bearing sori, 9"-36" tall.
  - Petiole (leaf stalk) smooth and pale green above, black at base, sparsely set with smooth, tan, ovate scales, 3½"-18". Petioles of fertile fronds much longer than those of sterile.
  - Blade lanceolate, 4"-16", lowest pinnae commonly somewhat shorter, blade tapering gradually to lobed tip.
  - Rachis (axis) green, slender, smooth.
  - Pinnae (leaflets) deeply cut to within 1mm of rib; lobes oblong, with smooth edge. About a dozen per frond, they are all perpendicular to the stem.
- Rootstalk long-creeping, slender (1mm-3mm), black.
  - Roots black, wiry, creeping, and shallow; typically few in number.

- Sori round, on underside of fronds near the midvein
  - indusia tan, often hairy
  - sporangia glabrous

Identification

- Identifiable by wetland habitat and frond characteristics.
- Distinguished from the closely related Long Beech Fern (*Phegopteris connectilis*), by its lowest leaflets growing perpendicular to the stem.
- Field Marks
  - marshy habitat
  - fronds emerge singly from horizontal rhizome as scattered fronds, no clumps
  - leafstalk longer than most of our ferns, with black at base
  - frond widest near base; all leaflets perpendicular to stem
  - 1½ times divided into tiny, more or less toothless lobes with forking veins.

Distribution

- Manitoba to Newfoundland, south to Texas, Louisiana, Mississippi, Alabama, Georgia, and Florida.
- Bermuda, Cuba, Peru; perhaps Mexico.
- Norway to Russia, south to Iberia, France, Italy, Greece, and the Ukraine.
- Armenia, Georgia, Japan, Kazakhstan, Kyrgyzstan, Russia, Tajikistan, Turkmenistan

Habitat

- Swamps, bogs, wet meadows, wet woods, and marshes, also along riverbanks and roadside ditches.
- Usually in rich, wet soil but not in standing water.

Associates

- *Trees:* Tamarack (*Larix larcina*), Black Spruce (*Picea mariana*), White Cedar (*Thuja occidentalis*)
- *Shrubs:* Speckled Alder (*Alnus incana rugosa*), Bog Birch (*Betula pumila*), Leatherleaf (*Chamaedaphne calyculata*)
- *Herbs:* Bluejoint Reed Grass (*Calamagrostis canadensis*), Sedges (*Carex oligosperma, Carex pauciflora*), Cottongrass (*Eriophorum vaginatum*), Orange Jewelweed (*Impatiens capensis*), White Panicled Aster (*Aster lanceolatus*), Purple Stem Aster (*Aster puniceus*), Parasol Aster (*Aster umbellatus*), Boneset (*Eupatorium perfoliatum*), Rough Bedstraw (*Galium asprellum*), Arrow-leaved Tearthumb (*Polygonum sagittatum*), and Sensitive Fern (*Onoclea sensibilis*).
- *Ground Covers:* Sphagnum Moss (*Sphagnum* spp.)

Reproduction

- By spore and vegetatively by rhizome

Propagation

- By spore or rhizome division

Cultivation

- Hardy to USDA Zone 3 (average minimum annual temperature -40°F)
- Cultural Requirements
  - Sun or shade
  - Soil slightly acidic, pH 5-7
  - Standing water to moist soil
- Can become invasive.
- Available by mail order from specialty suppliers.

POLYPODIUM VIRGINIANUM

- *Polypodium*, from the Greek, polus (*polus*), "many", podos (*podos*), "foot"; "many footed"
- *virginianum*, from the Latin, "of Virginia"

- Common Name, from the generic name
- Other common names include Rock Polypody, Rock Cap Fern, *Polypode de Virginie, Tripes-de-roches* (Qué), *Senikaladabagw* (Abenaki), *Ezodenda* (Jpn)

Taxonomy

- Kingdom *Plantae,* the Plants
- Division *Polypodiophyta,* the True Ferns
- Class *Filicopsida*
- Order *Polypodiales*
- Family *Polypodiaceae,* the Polypod Ferns
- Genus *Polypodium,* the Polypod Ferns
- Taxonomic Serial Number: 17242
- Also known as *Polypodium vulgare, Polypodium vulgare* var. *virginianum*

Description

- A small evergreen fern of rock crevices, 4"-12"
- Fronds spreading, compound with 10-20 alternate leaflets; lance-shaped; square base, pointed tip; 2"-10" long and 1¼"-2½" wide.
- Stem ungrooved; scattered, with thin light-brown scales or smooth
- Root System
- *Fruiting structure:* Large dot-like reddish-brown spore clusters in 2 rows on underside of leaves.
- Stems often whitish pruinose, slender, to 6 mm diam., acrid-tasting; scales weakly bicolored, lanceolate, contorted distally, base and margins light brown, sometimes with dark central stripe, margins denticulate.
- Leaves to 40 cm.
- Petiole slender, to 2 mm diam.

- Blade oblong to narrowly lanceolate, pinnatifid, usually widest near middle, occasionally at or near base, to 7 cm wide, somewhat leathery;
- rachis sparsely scaly to glabrescent abaxially, glabrous adaxially; scales lanceolate-ovate, usually more than 6 cells wide. Segments oblong, less than 8 mm wide; margins entire to crenulate; apex rounded to broadly acute; midrib glabrous adaxially. Venation free. Sori midway between margin and midrib to nearly marginal, less than 3 mm diam., circular when immature. Sporangiasters present, usually less than 40 per sorus, heads covered with glandular hairs. Spores more than 52 µm, tuberculate, surface projections more than 3 µm tall.

Identification

- Identifiable as
- Distinguished from
- Field Marks

Distribution

- Alaska to Newfoundland, south to South Dakota, Arkansas, Alabama, and Georgia.

Habitat

- Rocks, boulders, cliffs, and ledges
- Cliffs and rocky slopes; on a variety of substrates
- Mats on dry rocky outcrops in deciduous and coniferous forests.
- Rich woods and open woods; often on rocks or boulders

Associates

- Mosses and lichen

Reproduction

- Reproduces by spores and vegetatively by rhizomes

Propagation

- By rhizome division

Cultivation

- Hardy to USDA Zone 3 (average minimum annual temperature -40ºF)
- Available by mail order from specialty suppliers

ATHYRIUM FILIX-FEMINA

- *Athyrium* , from the Greek, *a*, "without", *qureos* (*thureos*), "shield"
- *filix-femina*, from the Latin, "fern-feminine"
- Common Name, an anglicized version of the Latin species name
- Other common names include Northern Lady Fern, *Athyrium Fougère-femelle* (Qué), *Skogburkne* (Nor), *Majbräken* (Swe), *Fjerbregne* (Dan), *Soreahiirenporras, Hiirenporras* (Fin), *Fjöllaufungur* (Is), *Wald-Frauenfarn, Gemeiner Waldfarn* (Ger), *Raineach Moire* (Gaelic)

Taxonomy

- Kingdom *Plantae*, the Plants
- Division *Polypodiophyta*, the True Ferns
- Class *Filicopsida*
- Order *Polypodiales*
- Family *Dryopteridaceae*
- Genus *Athyrium*, the Lady Ferns

Description

- A lacy, deciduous, densely clumping, perennial fern, 24"-36" tall.
- Fronds monomorphic, bright green, tufted, erect, 24"-36" long, 6-9" wide:
  - Petiole (leaf stalk) tan, reddish or brownish, scales brown to dark brown, linear-lanceolate.

- Blade elliptic, broadest near or just below middle, twice-cut
- Pinnae (primary leaflets) short-stalked or sessile, lanceolate
- Pinnules (secondary leaflets) deeply cut, linear to oblong

- Rootstalk stout, chaffy.
- Sori clustered at the pinnule base, straight, less frequently hooked or horseshoe-shaped; sporangial stalks bearing glandular hairs; indusia irregularly dentate, more or less ciliate. Ripen midsummer.

Identification

- Distinguished from Spinulose Woodfern (*Dryopteris carthusiana*) by its elongate, sometimes curved (rather than round) sori, which are covered by an indusium attached on one side.
- Field Marks
  - lacy, twice cut fronds
  - undersides of the leaf segments carry pale J-shaped sori

Distribution

- Circumpolar; Alaska to Labrador and Greenland, south in North America to Saskatchewan, Nebraska, Missouri, Illinois, Ohio, West Virginia, and North Carolina.

Habitat

- Meadows, open thickets, moist woods, and occasionally swamps.
- Can colonize cracks in rocks and crevices between rocks, as a pioneer species. More frequently occurs as a dominant on perennially wet soil with other herbs. Can survive severe battering if roots are protected and in constant contact with water.

- Major competitor in boreal and sub-boreal spruce forests. Commonly grows in the understory of White Spruce (*Picea glauca*) and Black Spruce (*Picea mariana*).

Fire

- Often occurs on wet sites that burn infrequently. Top-killed by fire, it resprouts from surviving rhizomes.

Associates

- *Trees:* Black Spruce (*Picea mariana*), White Spruce (*Picea glauca*)

History

- One of the most popular ferns during the Victorian fern craze.

Uses

- A common landscape and garden plant.

Reproduction

- Reproduces by spores and vegetatively by rhizomes

Propagation

- Division most successful method. Divide clumps in spring every few years and reposition crowns at soil level.
- Can be propagated from spores.

Cultivation

- A light green, fine textured, deciduous perennial; good for background foliage, naturalizing, woodland massing, and watersides. Easy to grow.
- Hardy to USDA Zone 2 (average minimum annual temperature -50ºF)
- Cultural Requirements
  - Full shade to partial shade or partial sun; full sun on wet sites
  - Rich, moist to wet, well-drained soil
  - Fertilization unnecessary

  - Growth rate moderate
  - 2'-3' H x 1'-2' W
  - Spacing 18"-24"
- Shelter from wind to protect fronds from breaking.
- A highly variable species, with numerous varieties taken into cultivation, some extremely odd in appearance. More than 300 cultivars have made their way to the market.
- Available by mail order from specialty suppliers or at local nurseries.

ONOCLEA SENSIBILIS

- *Onoclea,* from the Greek onos *(onos)*, "vessel," and kleio (*kleio*), "to close", referring to the closely rolled fertile fronds.
- *sensibilis,* from the Latin, "sensitive"
- Common Name, from the observation of early settlers that it was very sensitive to frost, the fronds dying quickly when first touched by frost.
- Other common names include Bead Fern, Meadow Brake, *Onoclée Sensible* (Qué)

Taxonomy

- Kingdom *Plantae,* the Plants
- Division *Polypodiophyta,* the True Ferns
- Class *Filicopsida*
- Order *Polypodiales*
- Family *Dryopteridaceae*
- Genus *Onoclea*
- Taxonomic Serial Number: 17636
- Also known as *Angiopteris sensibilis, Calypterium sensibile, Onoclea augescens, Onoclea interrupta, Onoclea obtusilobata, Ragiopteris obtusilobata, Ragiopteris onocleoides, Riedlea sensibilis*

Description

- A , , ¾, °, é
- A deciduous, coarse textured, perennial fern, with broader leaves and pinnae (leaflets) than most other North Country ferns; 18"-24" tall.
- Sterile Fronds light green, leathery, broad, and almost triangular, typically tilted up and back.
  - Petiole (leaf stalk) brittle, smooth, and usually longer than blade with a shallow furrow in front; yellow/ pale tan, and dark brown at base with a few scales.
  - Blade of 8-12 pairs of nearly opposite leaflets with wavy margins, prominent network of veins, and sparse white hairs on underside.
  - Rachis (axis) winged, more broadly toward tip.
  - Pinnae (primary leaflets) lanceolate to oblong, with or without teeth or with a few undulations. Lower leaflets long and tapered at both ends; upper leaflets with little or no tapering toward rachis.
  - Pinnules (secondary leaflets) blades mostly 4"-12" long, 4"-14" wide, with 8-12 pairs of opposite pinnae, these sinuate to pinnatifid, ½"-1¼" wide, sparsely white-hairy on the veins beneath; petioles shorter to about as long as the blade.Sterile leaves yellow-green, deltate, coarsely divided, 5"-13½" × 6"-12". Petiole of sterile leaf black, 22-58 cm, flattened at base; rachis winged, becoming broader toward apex. Pinnae 5-11 per side, lanceolate; proximal pinnae 9-18 cm, margins entire, sinuate, or laciniate.
- Fertile Fronds brown, shorter than sterile leaves (approximately 1' tall), and structurally unlike the green sterile fronds. Fertile fronds produced Aug-Sep, often persistent into the following year.
  - Petiole (leaf stalk)Petiole 19-40 cm, base sparsely scaly.
  - Blade

- Rachis (axis)
- Pinnae (primary leaflets) numerous, compact, upward-pointing Pinnae linear, 5-11 per side, 2.5-5 cm; pinnae strongly ascending, 2-5 cm long, Sporophyll leaves green, becoming black at maturity, oblong, 7-17 × 1-4 cm.
- Pinnules (secondary leaflets) many hardened, beadlike subleaflets that become dark brown when mature. ultimate segments revolute to form beadlike structures, 2-4 mm diam.divided into bead-like pinnules with inrolled margins enclosing the sori, the pinnules 3-4 mm wide, becoming dry, hard, eventually separating to release the spores;
- Sensitive fern is easily killed with the first frost, leaving behind the stiff, beaded fertile stalks. blade pinnate-pinnatifid, mostly 5-15 cm long;

• Root System of creeping rhizomes 4mm-7mm thick, growing near the soil surface; stout, brown, smooth, and extensively branched and spreading. Numerous roots grow along the rhizomes and produce a fibrous mat.

• Fiddleheads The curled leaves (fiddleheads) emerging from rhizomes form a distinctive, pale red mass in the spring

• Fruit spore cases produced within the hardened, beadlike sections of the fertile leaflets, becoming dark brown at maturity. sori globose, covered by a delicate, hoodlike indusium.

- Sori clustered like beads or grapes on the upright fertile fronds
- Spores minute, on separate fertile fronds, within bead-like modified leaflets

• Leaf forms with pinnae intermediate between those of sporophylls and sterile leaves, or with pinnae fertile only on one side of the blade, can occur on plants that also bear normal leaf forms.

- As with *Matteuccia struthiopteris,* sporophylls of *Onoclea sensibilis* persist through the winter and release the green spores in spring before the sterile leaves expand.

Identification

- Identifiable as
- Distinguished from
- Field Marks
  - large, deeply pinnatifid fronds
  - spherical spore-bearing bodies borne on a separate stalk

Distribution

- Manitoba to Newfoundland, south to Texas, the Gulf Coast, and Florida
- Also, East Asia
- Naturalized in western Europe

Habitat

- Wet meadows, thickets, and woods; stream and riverbanks; swamps and bogs; usually in slightly acidic soil.
- In sunny or shaded locations, often forming thick stands; uncommon in forested environments.

Uses

- Implicated in the poisoning and death of horses grazing low, wet areas.

Reproduction

- Reproduces by spores and vegetatively by rhizomes.

Propagation

- By spores or rhizome division in spring.

Cultivation

- A low maintenance fern for moist sites which, despite its name, tolerates the toughest of conditions.

- Hardy to USDA Zone 3 (average minimum annual temperature -40°F)
- Cultural Requirements
  - Shade or part shade; will tolerate sun with adequate moisture
  - Average garden soil, on the acidic side
  - Moisture: moist soil; will tolerate wet soils and can be used near water.
  - Fertilization unnecessary
- Size 18"-24"W × 12"-24"H
- Good for bogs, natural gardens, woodland drifts, shade groundcover
- Available by mail order from specialty suppliers or at local nurseries
- Spreads to form colonies; can become weedy and invasive
- Winter survival improves if dried fronds are left on plant over winter.

DRYOPTERIS FRAGRANS

- *Dryopteris*, from the Greek, drus (*drys*), "oak", pteris (*pteris*), "fern", "fern of the oak wood"
- *fragrans*, from the Latin, "fragrant"
- Common name from the aroma of the fronds when handled.
- Other common names include Fragrant Wood Fern, Fragrant Shield Fern, Fragrant Cliff Fern, *Dryoptère Fragrante, Dryoptère Odorante* (Qué), *Tuoksualvejuuri* (Fin)

Taxonomy

- Kingdom *Plantae*, the Plants
- Division *Polypodiophyta*, the True Ferns
- Class *Filicopsida*
- Order *Polypodiales*

- Family *Dryopteridaceae,* the Wood Ferns
- Genus *Dryopteris,* the Wood Ferns
- Taxonomic Serial Number: 17536
- Also known as *Polypodium fragrans, Thelypteris fragrans*
- A northern species not closely related to North American *Dryopteris*. The only known hybrid is with *Dryopteris marginalis*, producing *Dryopteris* × *algonquinensis*.

Description

- An atypical "Wood Fern" of rock faces.
- Fronds monomorphic, leathery, evergreen, strongly tapered at the base, and ½"-2½"× 2½"-16"; old leaves persistent as conspicuous grey or brown clump at plant base.
  - Petiole (leaf stalk) ¾"-4½" long, to 1/3 length of leaf, scaly throughout; scales dense, brown to red-brown.
  - Blades green, linear-lanceolate, and twice-cut; upper surface smooth, underside densely scaly; aromatic when handled.
  - Pinnae (primary leaflets) more or less in plane of blade, linear-oblong, and densely crowded, often overlapping and inrolled; lowest pairs much reduced.
  - Pinnules (secondary leaflets) with toothed edges; lowest pair longer than adjacent pinnules.
- Rootstalk short, thick, and erect; covered with brown scales.
  - Roots grey or black, sparse.
- Sori large, midway between midvein and edge of leaf, with distinct indusium.
  - Indusium ovate, whitish, and often overlapping, becoming brown with ragged margins.

Identification

- Identifiable by its rocky habitat, small size, and persistent leaf bases.

- Distinguished from other *Dryopteris* species by looking nothing at all like them, from Common Polypody (*Polypodium virginiana*) by having twice cut fronds, and from the Cliff Ferns (*Woodsia* species) by its very densely crowded leaflets, its frond tapered at the base, and its short leaf stalk.
- Field Marks
  - rocky habitat
  - twice-cut fronds strongly tapered at base
  - densely crowded leaflets
  - persistent old leaves at base of plant

Distribution

- Alaska to Newfoundland and Greenland, south to British Columbia, Alberta, Saskatchewan, Manitoba, Minnesota, Wisconsin, Michigan, New York, Vermont, New Hampshire, and Maine.
- Also northern Europe (Estonia and Finland) and Asia (Siberia and Japan).

Habitat

- Shaded cliffs, talus, and scree; in crevices and on rock.
- Dry, moderately well drained rock, gravel, and till, with low organic content.

Fire

Associates

- *Shrubs:* Juneberry (*Amelanchier* spp.), Low Bush Honeysuckle (*Diervilla lonicera*)
- *Herbs:* Columbine (*Aquilegia canadensis*), Harebell (*Campanula rotundifolia*), Pale Corydalis (*Corydalis sempervirens*), Fringed Bindweed (*Polygonum cilinode*)
- *Ferns:* Common Polypody (*Polypodium virginianum*), Rusty woodsia (*Woodsia ilvensis*)
- *Ground Covers:* Rock Spikemoss (*Selaginella rupestris*)

Reproduction

- By spore and vegetatively by rhizome.

Propagation

- By rhizome division.

Cultivation

- Hardy to USDA Zone 2 (average minimum annual temperature -50ºF)
- Cultural Requirements
  - Sun to part shade
  - Well-drained, rocky soil, with low organic content
  - Dry to medium moisture
- A small fern suitable for alpine rock gardens.
- Occasionally available by mail order from specialty suppliers.

MATTEUCCIA STRUTHIOPTERIS

- *Matteuccia*, named in honor of Carlo Matteucci (1800-1868), an Italian physicist.
- *struthiopteris*, from the Greek, strouqeios (*stroutheios*), "of an ostrich", and pteris (*pteris*), "fern"
- Common Name, from the resemblance of the fronds to the plumes of the large flightless bird of Africa.
- Other common names include Fiddlehead Fern, Garden Fern, Hardy Fern, *Fougère-à-l'autruche* (Qué), *Strutbräken, Foderbräken* (Swe), *Strutsveng (Nor), Strudsvinge (Dan), Kotkansiipi (Fin), Straußfarn (Ger), Matteuccia* (It), *Struccpáfrány* (Hun), *Pióropusznik strusi* (Pol)

Taxonomy

- Kingdom *Plantae*, the Plants
- Division *Polypodiophyta*, the True Ferns
- Class *Filicopsida*

- Order *Polypodiales*
- Family *Dryopteridaceae*
- Genus *Matteuccia*
- Taxonomic Serial Number: 17596
- Also known as *Matteuccia pensylvanica, Matteuccia struthiopteris* var. *pensylvanica, Matteuccia struthiopteris* var. *pubescens, Onoclea struthiopteris, Onoclea struthiopteris* var. *pensylvanica, Pteretis nodulosa, Pteretis pensylvanica.*

Description

- A large, feathery, deciduous fern, 3'-5' tall.
- Sterile Frond large, green, oblanceolate, up to 60" long, 12" wide, forming symmetrical, vaselike cluster.
  - Petiole (leaf stalk) black, to 12", flattened at base, becoming deeply grooved, scales pale orange-brown.
  - Pinnae (primary leaflets) linear, 20-60 per side, the longest near the tip, gradually decreasing in length toward base; leaflets further subdivided into 20-40 segments.
- Fertile Frond dense and rigid, arising from the center of the clump in mid to late summer, persisting through winter; green maturing to dark brown.
  - Petiole (leaf stalk) 3"-8", with scaly base.
  - Blade lyre-shaped, oblong to oblanceolate, 6"-16"× 1"-2½".
  - Pinnae (primary leaflets) linear, 30-45 per side, 1¼"-2¼" long, rolled inward to clasp spores, forming hard "pod".
- Stem stout, green, covered with white hairs.
- Rhizome large, with a braided appearance.
- Roots black and wiry.
- As with Sensitive Fern (*Onoclea sensibilis*), forms intermediate between sterile fronds and fertile fronds are sometimes found.

Identification

- Distinguished from other large ferns by the distinctive fertile frond, when present. In the absence of the spore-bearing structure, the twice-cut sterile fronds with white hairs on the stems mark this species.
- Field Marks
  - large size
  - distinctive fertile frond, utterly unlike sterile fronds
  - twice-cut sterile fronds, broad near tip and tapering gradually to base
  - fine white hairs on stem

Distribution

- Alaska to Newfoundland, south to British Columbia, Alberta, Saskatchewan, North Dakota, Missouri, Illinois, Indiana, Ohio, West Virginia, and Virginia.
- Also Scandinavia, Central Europe, Russia, and Asia; introduced into Ireland and Great Britain from the continent.

Habitat

- Moist soil in deciduous and mixed forest, wooded river bottoms, and swamps.
- Often in alluvial or mucky swamp soils.

Associates

- *Trees:* Green Ash (*Fraxinus pennsylvanica*), Balsam Poplar (*Populus balsamifera*), Bur Oak (*Quercus macrocarpa*)

History

- The edible fiddlehead is the state vegetable of Vermont.

Uses

- The most commonly sold species of generic garden fern.
- Grown commercially for the decorative fronds.

- Fiddleheads (young coiled sterile fronds) are considered a delicacy. Collected in early spring, they support a local canning industry in New England and adjacent Canada. For more information on preparation of fiddleheads, see University of Maine Extension Bulletin #4198.

Reproduction

- Reproduces by spores and vegetatively by rhizomes.
- Fertile fronds produced after vegetative fronds and persist throughout the following winter. Spores shed mid-winter.

Propagation

- Division most successful method. Best transplanted when dormant in early spring or fall but can, with care, survive transplantation at any time.
- Can be grown from spores with sufficient patience.

Cultivation

- An elegant, robust garden fern for moist, shaded sites. Very easy; low maintenance.
- Hardy to USDA Zone 3 (average minimum annual temperature -40°F)
- Cultural Requirements
  - Light to full shade. Will yellow and burn in hot sun.
  - Soil highly organic, moist to swampy or boggy, slightly acid (pH of 5-6.5). Does well in ordinary garden loam or clay, however.
  - Consistent moisture; should not be allowed to dry out between waterings. The greater the sun exposure, the greater the moisture requirement.
  - Spacing: 24"-36"
  - Fertilization unnecessary
- Size 24"W × 36"-60"H, the more sun and moisture, the larger the plant.

- Growth rate moderate. However, under good conditions tends to spread aggressively by stout rhizomes. Excercise caution near smaller, less robust plants.
- Good for foliage backdrop, foundation plantings, naturalizing.
- Readily available by mail order or at local nurseries

DRYOPTERIS CRISTATA

- *Dryopteris*, from the Greek, drus (*drys*), "oak" and pteris *(pteris)*, "fern", "fern of the oak wood"
- *cristata*, from the Latin *cristatus*, "tufted, crested"
- Common name from
- Other common names include Crested Shield Fern, Crested Buckler Fern, Buckler Fern, Narrow Swamp Fern, *Dryoptère à Crêtes* (Qué), Crested Buckler-fern, Gray Crested Shield Fern (UK), *Granbräken* (Swe), *Vasstelg* (Nor), *Butfinnet Mangeløv* (Dan), *Korpialvejuuri* (Fin), *Kammfarn* (Ger), *Tarajos pajzsika* (Hun)

Taxonomy

- Kingdom *Plantae,* the Plants
- Division *Polypodiophyta,* the True Ferns
- Class *Filicopsida*
- Order *Polypodiales*
- Family *Dryopteridaceae,* the Wood Ferns
- Genus *Dryopteris*, the Wood Ferns
- Also known as *Aspidium cristatum, Lastrea cristata, Nephrodium cristatum, Polypodium cristatum, Polystichum cristatum, Thelypteris cristata*
- Thought to have originated from a cross of *Dryopteris ludoviciana* and the hypothetical, and presumably extinct, *Dryopteris semicristata.*
- Hybridizes with five species, producing plants identifiable by narrow blades and triangular lower pinnae.

Description

- A gaunt fern of wetlands.
- Fronds dimorphic; fertile leaves deciduous, 10"-24"; sterile leaves smaller, 6"-12", forming persistent, evergreen rosette.
  - Petiole (leaf stalk) 1/4-1/3 length of leaf, scaly at least at base; scales scattered and tan.
  - Blade green, narrowly lanceolate or with parallel sides, twice-cut.
  - Rachis (axis) green; slightly scaly toward base.
  - Pinnae (primary leaflets) of fertile leaves triangular, twisted out of plane of blade and perpendicular to it; lower pair somewhat reduced in size.
  - Pinnules (secondary leaflets) uncut, with spiny teeth; lowest pair of pinnules longer than adjacent pinnules; basal basiscopic pinnule and basal acroscopic pinnule equal.
- Rootstalk dark brown, stout, creeping or ascending, with numerous brown scales.
  - Roots black, wiry, and widely spreading; numerous.
- Sori midway between midvein and leaf edge.
  - Indusia kidney shaped, with minute hairs.

Identification

- The narrow frond with short, widely spaced, and tilted leaflets is unlike anything else in the North Country.
- Distinguished from other *Dryopteris* species by habitat and narrow fronds with widely spaced leaflets tilted out of the plane of the blade.
- Field Marks
  - wetland habitat
  - narrow fronds with widely spaced, tilted leaflets

Distribution

- Circumboreal, British Columbia to Newfoundland, south to NE Washington, Idaho, NW Montana, Alberta, Saskatchewan, Manitoba, Minnesota, Iowa, Illinois, Indiana, Ohio, West Virginia, and North Carolina.
- Europe from Norway to Russia, south to the UK, Spain, Italy, and Romania.

Habitat

- Moist woods, thickets, marshes, swamps, sphagnum bogs, and open shrubby wetlands.

Reproduction

- By spore and vegetatively by rhizome.

Propagation

- By rhizome division.

Cultivation

- Hardy to USDA Zone 3 (average minimum annual temperature -40°F)
- Cultural Requirements
- Sun to part shade.
- Moist to wet soil
- Occasionally available by mail order from specialty suppliers.

4

Dryopteris

SCIENTIFIC CLASSIFICATION

Kingdom : Plantae

Division : Pteridophyta

Class : Pteridopsida

Order : Dryopteridales

Family : Dryopteridaceae

Genus : *Dryopteris*

Adans

SPECIES

D. goldiana, Goldie's Fern.

Dryopteris (commonly called Wood Ferns, Male Ferns and Buckler Ferns) is a genus of about 250 species of ferns with distribution in the temperate Northern Hemisphere, with the highest species diversity in eastern Asia. Many of the species have stout, slowly creeping rootstocks that form a crown, with a vase-like ring of fronds. The sori are round, with a peltate indusium. The stipes have prominent scales.

Hybridization is a well-known phenomenon within this group, with many species formed by hybridization.

Dryopteris species are used as food plants by the larvae of some Lepidoptera species including *Batrachedra sophroniella* (which feeds exclusively on *D. cyatheoides*) and *Sthenopis auratus*.

SELECTED SPECIES

Dryopteris abbreviata

Dryopteris aemula - Hay-scented Buckler Fern

Dryopteris affinis - Scaly Male Fern

Dryopteris alpestris

Dryopteris amurensis

Dryopteris apicisora

Dryopteris arguta - Coastal Wood Fern

Dryopteris azorica

Dryopteris barbigera

Dryopteris bissetiana

Dryopteris blandfordii

Dryopteris bodinieri

Dryopteris borreri

Dryopteris campyloptera - Mountain Wood Fern

Dryopteris canaliculata

Dryopteris carthusiana - Narrow Buckler Fern

Dryopteris caucasica

Dryopteris caudifrons

Dryopteris celsa - Log Fern

Dryopteris championi

Dryopteris changii

Dryopteris chapaensis

Dryopteris chinensis

Dryopteris chrysocoma

Dryopteris cinnamomea - Cinnamon Wood Fern

Dryopteris clintoniana - Clinton's Wood Fern

Dryopteris confertipinna

Dryopteris cordipinna

Dryopteris corleyi costalisora

Dryopteris crassirhizoma

Dryopteris crispifolia

Dryopteris cristata - Crested Buckler Fern

Dryopteris cyatheoides

Dryopteris cycadina

Dryopteris cyclopeltiformis

Dryopteris dehuaensis

Dryopteris dickinsii

Dryopteris dilatata - Broad Buckler Fern

Dryopteris discrita

Dryopteris enneaphylla

Dryopteris erythrosa

Dryopteris erythrosora - Autumn Fern

Dryopteris expansa - Northern Buckler Fern

Dryopteris fibrilosa

Dryopteris fibrilosissima

Dryopteris filix-mas - Male Fern

Dryopteris floridana

Dryopteris formosana

Dryopteris fragrans - Fragrant Buckler Fern

Dryopteris fructuosa

Dryopteris fuscipes

Dryopteris gamblei

Dryopteris goerigiana

Dryopteris goldiana - Goldie's Wood Fern

Dryopteris gongboensis

Dryopteris guanchica

Dryopteris gushanica

Dryopteris gushiangensis

Dryopteris gymnophylla

Dryopteris gymnosora

Dryopteris hendersoni

Dryopteris hexagonoptera

Dryopteris hirtipes

Dryopteris huanganshanensis

Dryopteris hupehensis

Dryopteris hwangii

Dryopteris hypophlebia

Dryopteris immixta

Dryopteris incisolobata

Dryopteris integriloba

Dryopteris integripinnula

Dryopteris intermedia - Intermediate Wood Fern

Dryopteris junlianensis

Dryopteris juxtaposita

Dryopteris kinkiensis

Dryopteris labordei

Dryopteris lacera

Dryopteris lancipinnula

Dryopteris lepidopoda

Dryopteris lepidorachis

Dryopteris liyangensis

Dryopteris ludoviciana - Southern Wood Fern

Dryopteris manshurica

Dryopteris marginalis - Marginal Wood Fern

Dryopteris matsumurae

Dryopteris minjiangensis

Dryopteris monticola

Dryopteris montigena

Dryopteris nanpingensis

Dryopteris neolepidopoda

Dryopteris nigrosquamosa

Dryopteris nyalamensis

Dryopteris nyingchiensis

Dryopteris odontoloma

Dryopteris oreades - Mountain Male Fern

Dryopteris pacifica

Dryopteris pallida

Dryopteris paludicda

Dryopteris panda

Dryopteris parasparsa

Dryopteris pedata

Dryopteris peninsulae

Dryopteris podophylla

Dryopteris polita

Dryopteris prosa

Dryopteris pseudo-sikkimensis

Dryopteris pseudoatrata

Dryopteris pseudodontoloma

Dryopteris pseudofibrillosa

Dryopteris pseudomarginata

Dryopteris pseudouniformis

Dryopteris pulcherrima

Dryopteris qandoensis

Dryopteris quatanensis

Dryopteris reflexosquamata

Dryopteris remota

Dryopteris rigida

Dryopteris rosthornii

Dryopteris sacrosancta

Dryopteris sanmingensis

Dryopteris saxifraga scottii

Dryopteris semipinnata

Dryopteris sericea

Dryopteris serrato-dentata

Dryopteris shensicola

Dryopteris sichotensis

Dryopteris sieboldii sikkimensis

Dryopteris silaensis

Dryopteris sino-sparsa

Dryopteris sinoerythrosora

Dryopteris sinofibrillosa

Dryopteris sordidipes

Dryopteris sparsa

Dryopteris spinulosa

Dryopteris squamifera

Dryopteris squamiseta

Dryopteris stenolepis

Dryopteris subatrata

Dryopteris subbarbigera

Dryopteris subexaltata

Dryopteris sublacera

Dryopteris sublaeta

Dryopteris submarginata

Dryopteris submontana - Rigid Buckler Fern

Dryopteris subtriangularis

Dryopteris tarningensis

Dryopteris tenuicula tenuissima

Dryopteris thibetica

Dryopteris tieluensis

Dryopteris tsangpoensis

Dryopteris tyrrhena

Dryopteris uniformis

Dryopteris varia venosa

Dryopteris villarii

Dryopteris wallichiana

Dryopteris wenchuanensis

Dryopteris wuyishanensis

Dryopteris yigongensis

Dryopteris yungtzeensis

Dryopteris zayuensis

CULTIVATION AND USES

Many *Dryopteris* species are highly desired as garden ornamental plants, especially *D. erythrosa* (autumn fern, often sold in garden outlets) and *D. filix-mas,* a very popular garden fern in the British Isles and Europe, with numerous cultivars.

Dryopteris filix-mas was throughout much of recent human history widely used as a vermifuge, and was the only fern listed in the U.S.

5

Growing Ferns

INTRODUCTION

If you find yourself fascinated with ferns, congratulate yourself. You are among a select group of people described as follows by Herbert Stansfield in the mid 1800s: "The bright colors in flowers are admired by the least intellectual, but the beauty of form and textures requires a higher degree of mental perception and more intellect for its proper appreciation."

Growing ferns indoors first became popular in England in the 1800s, and reached its peak in the 1850s, during the so-called Victorian Fern Craze. Fern cultivation was a highly fashionable fad, which arose purely for aesthetic reasons. Fortunately, technological developments in heating and the manufacture of glass coincided, allowing people to pursue their new passion. The intense preoccupation with ferns further stimulated academic studies, and scientific interest was also fostered by improvements in microscopes and their greater availability. As in other sciences, in fashion, music, dance, and the other arts, horticultural trends cycle around, coming in and going out of style. Presently these "semi-antique" plants are enjoying a revival, much like heirloom vegetable and flower varieties.

Nephrolepis exaltata, Boston fern.

To most people, the word "fern" conjures up certain images—delicate, lacy, airy greenery luxuriating in a shady little spot. While this may be true for some ferns, these plants are more diverse than most people think. Ferns originated about 300 million years ago, and there are about 12,000 species of ferns and fern allies growing on the earth today. Their leaves vary in size from 1/8 inch to 60 feet, and their colors range through the full spectrum of green and also white, silver, golden yellow, red, pink, copper, and burgundy, even blue. Fern fronds come in different textures and can be thick and leathery, somewhat succulent, hairy, waxy, or super thin—only a single cell layer thick! The majority of ferns make their homes in moist tropical forests, but they also venture into cold temperate zones, bodies of water, and even the desert. These somewhat primitive plants do not flower or set seed, but rather reproduce by spores. However, they do have a vascular system, a relatively advanced feature, which is basically a network of veins for the transport of water, nutrients, and food. Their stems are often modified into "rhizomes"; at times these are very obvious and showy, and at other times they're hidden in the soil.

From a design perspective, ferns are very versatile. Their quiet, graceful beauty lends itself equally well to classical, formal styles and rustic, informal settings. A perfect example is the Boston fern, the most common fern grown in America today.

Most people shy away from growing ferns, thinking that they are all too difficult. However, with proper selection and care, they are very gratifying plants.

HUMIDITY

With many ferns, humidity is the key to success. Most require at least 35 percent; 40 to 50 per cent is even better. The typical home has 10 to 25 percent, especially in winter, and in summer as well if an air conditioning system is being used.

Humidity is a relative factor, hence the term "relative humidity" (RH). Cooler air has less of a capacity for holding moisture than warmer air. With the same amount of actual moisture in the air, 55° F. will have a higher relative humidity reading than 65° F. So you begin to see why keeping plants cooler, especially in winter, is beneficial. There are several things you can do to increase the humidity around your plants. Running a humidifier will produce the best results, and it's also good for you, your wood furniture, books, paintings, and so on. Mass or group your plants together, and/or set them on trays of gravel filled with water, but be certain to raise the bottom of your containers above the water level. Add a few pieces of charcoal to the tray, and clean it thoroughly with a 10-per cent bleach solution or scalding water every two months or so.

Victorian growth chambers were needed to provide humidity, but also to maintain warmth and protect plants from the noxious fumes of coal heating.

You can also mist ferns with larger leaves, but bear in mind that one little spritzing a day won't increase humidity for very long—when the droplets have dried up, the humidity is basically gone! An alternative is to enclose your plants in a terrarium of some kind. Fancy growth chambers date back to Victorian England. At the time they were needed not only to provide humidity, but also to maintain warmth and protect plants from the fumes of the burning coal that was used as a heating fuel. Now such growth chambers are sometimes used to provide additional warmth, but mainly to maintain humidity.

TEMPERATURE

The majority of ferns that are available to us as houseplants originate in the tropics and prefer a daytime temperature ranging from 65 to 75° F.; the nighttime temperature should be 10 degrees cooler, ranging from 55 to 65° F.

Light

Contrary to what many people think, ferns do not grow in the dark! They are photosynthetic creatures just like other plants. Ideally, ferns prefer bright, indirect light for best growth, and many will benefit from a little sun in winter. Many will tolerate lower light, but under these conditions, they merely survive and won't grow very much.

Soil Mixes

When it comes to potting and soil mixes, it is useful to divide ferns into two categories: terrestrial and epiphytic. Terrestrial ferns grow naturally in the ground and make good potted plants. For these ferns, use a soft, organic soil mix such as: 2 parts loam, 2 parts organic matter (leaf mold, compost, or peat moss), 1 part perlite, 1 part vermiculite, 1 part charcoal, and 1 part very fine grade orchid bark. Ferns tend to have small, delicate root systems, so when potting, don't disturb the root ball too much and firm the soil in place, but don't "pack" it in.

Epiphytes naturally grow up in trees, or on rocks, etc., not in the earth. Most epiphytic ferns have an extensive rhizome system that grows around their "support structure." The rhizomes will grow all around your basket or pot, so they should be "planted" on the surface and then allowed to grow "outside" the container. Epiphytes require a different mix: 1 part of the terrestrial mix above combined with ½ part of very fine grade orchid bark and ½ part tree fern fiber.

Pests

The most common pests on ferns are scale and mealybug. Picking mealybug off by hand or with a Q-tip dipped in alcohol or soapy water is very effective, although tedious. Scale is even more difficult to remove manually. Another non-chemical control is to release some beneficial, predatory insects.

Propagation

Propagating many ferns is easy. Those that grow in a clump can be divided using typical methods. Many epiphytes can be propagated from rhizome (stem) cuttings. These take best if the pieces have at least one leaf and preferably roots. Some ferns, commonly called "mother ferns," make baby plantlets along their leaves, and these can easily be separated and grown in a separate pot. Most ferns can also be grown from spores, which is not as difficult as it may sound, but takes much longer.

FERNS FOR INDOORS

- Rough Maidenhair *Adiantum hispidulum*—A terrestrial clump-forming fern 8 to 12 inches tall, this species is not as delicate as the common maidenhair, but has an interesting frond shape. Young leaves emerge copper-colored and mature to green.

- Mother Fern *Asplenium bulbiferum*—This terrestrial, clump-forming fern has medium green, somewhat fleshy, dissected leaves. One of the "mother ferns," it produces baby plantlets on its leaves, making it easy to propagate.

- Bird's Nest Fern *Asplenium nidus*—Naturally an epiphyte, this fern can grow up to 3 feet (usually 1 foot). The leathery, strap-shaped fronds are bright green and form a nice rosette.

- Rochford Holly Fern *Cyrtomium falcatum* 'Rochfordianum'—This terrestrial clump former grows to about 1 foot. Decorative, glossy, dark green leaves have toothed edges resembling holly.

- Rabbit's Foot Fern *Davallia species*—These epiphytic ferns have fronds 1 to 2 feet long and fuzzy rhizomes. Their leaves are usually very finely dissected. Excellent for hanging baskets.

- Squirrel's or Bear's Foot Fern *Humata tyermannii*—This epiphyte strongly resembles *Davallia,* but it is smaller, with fronds 8 to 12 inches, and has more slender, whitish colored rhizomes. It might be a little easier to grow.

- Japanese Climbing Fern *Lygodium japonicum*—A terrestrial clump-forming fern with twining stems climbing 5 feet (to 20 feet). Wiry stalks support hand-shaped segments. Cut down old growth to avoid a large, tangled mass.

- Sword Fern *Nephrolepis cordifolia*—This clump former can be grown terrestrially or epiphytically, but with erect fronds it is best as a potted plant. This species is more sturdy and tolerant of low light than the common Boston fern type, *Nephrolepsis exaltata.*

- Button Fern *Pellaea rotundifolia*—This terrestrial fern forms clumps with leaves 8 inches long. It is a charming potted plant with very dark green, somewhat glossy, small, rounded leaflets resembling buttons.

- Hare's Foot Fern *Phlebodium species*—An epiphyte, this plant can be grown potted or in a hanging basket. Fronds are variable from 1 to 4 feet long and bright green to steel blue-gray. The stout creeping rhizomes are covered with orange-coloured scales.

- Staghorn Fern *Platycerium species*—This epiphyte is best grown in a basket or mounted on cork or wood slabs. These bizarre, curious plants resemble stag's antlers and are very interesting and decorative.
- Tsus-sima Holly Fern *Polystichum tsus-simense*—A terrestrial clump former 8 to 12 inches high. Its very attractive stiff, leathery, somewhat glossy dark green leaves have black veins and bristly tips.
- Whisk Fern *Psilotum nudum*—Naturally terrestrial and epiphytic, this plant, which is not a true fern, grows in a clump 6 to 18 inches high. A curiosity, this primitive plant has no true roots or leaves; it consists of a bunch of green, forking stems.
- Brake Fern *Pteris species*—Terrestrial clump formers, members of this genus are very variable in size (8 to 24 inches) and leaf shape, which is usually nicely dissected. Some are variegated with white or silver. The fertile (spore-bearing) fronds are distinct in shape. The very common Cretan brake prefers lime in the soil as do some others.
- Leather Leaf *Rumohra adiantiformis*—A terrestrial clump-forming species that grows 1 to 1½ feet. It has very shiny dark green, leathery leaves.
- Tree Fern *Cyathea* (*Sphaeropteris*) and other tree ferns—Tree ferns are palm-like in habit. They are elegant but delicate and must not dry out.

CHARACTERISTICS AND STRUCTURES OF PLANTS

Photosynthesis - Plants obtain their energy by converting light energy into chemical energy by means of photosynthesis, which involves two basic sets of reactions that take place in the chloroplast. The light dependent reactions involve the absorption of light energy by specialized pigments including chlorophyll, and the reaction within the thylakoid. The energy is absorbed by electrons and used to create ATP and a compound called NADPH provides the hydrogen for the subsequent reaction. These compounds carry energy and hydrogen ions (in the case of NADPH) into the light independent reactions. These reactions bring carbon dioxide into the plant's metabolism. Eventually, all these materials are combined into glucose and other carbohydrates. In general, the plant uses its root to extract water in the soil, its leaf to absorb carbon dioxide from the air, energy from the sun, and the stem for transportation of water and nutrients). Oxygen is a by product exits through the leaf.

Leaves are the organs of photosynthesis in vascular plants. A leaf usually consists of a flattened blade and a petiole, which connects the blade to the stem. The blade may be single or it may be composed of several leaflets. Externally, it is possible to see the pattern of the leaf veins that bring water to the leaf; they are the final extensions of the vascular tissue. At the top and bottom of the leaf is a layer of epidermal tissue that often bears protective hairs and glands that produce irritating substances to discourage herbivores. The epidermis is covered by a waxy cuticle that keeps the leaf from drying out. Unfortunately, it also prevents gas exchange because the cuticle is not gas permeable. However, the epidermis, particularly the lower one, contains openings called stomata that allow gases to move into and out of the leaf. Each stoma has two guard cells that regulate its opening and closing. The body of a leaf is composed of mesophyll tissue, which contain many chloroplasts and carry on most of the photo-synthesis for the plant. Leaf veins consist of a strand of xylem and a strand of phloem surrounded by a bundle sheath. The bundle sheath cells in C3 plants do not contain chloro-plasts; while the C4 plants characteristically have chloroplasts and are more efficient in fixing carbon dioxide.

Locomotion: It is often said that since plants simply transform light into chemical energy, they need only to be anchored firmly in one place with a maximum surface area to capture sunlight. There is no need for plants to move or to have sophisticated nervous systems. Actually, some of them have quite elaborate and creative ways of moving their leaves in response to a wide variety of stimuli, such as touch and light. Leaf movements can be as fast as a millisecond in the Venus flytrap or as slow as a half-hour in sun-tracking plants. Motor cells located in the region—called the pulvinus—where the leaf connects to the stem, control the movement. These cells either shrink or swell because of the influx or efflux of water. For example, if the cells at the top of the pulvinus swell and the bottom cells shrink, the leaf tilts downward. The changes in cell volume are a direct result of changes in the concentration of certain ions in the cell, such as potassium and chloride. For example, when a large amount of potassium enters the cell through special potassium channels, a large amount of water must enter so that the concentration of potassium stays constant. It is unclear what is the signaling mechanism to open the ion channels. Different plants have evolved different signaling mechanisms depending on what triggers the leaves to move.

Reproduction and Growth: Asexual reproduction is very common among plants, while sexuality is not well-defined (90% have both male and female parts). Plant cells are rarely terminally differentiated. Since most cells are totipotent, detached limb can be re-generated readily. Plants do not establish a germ line (special cells destined to produce gametes). Higher plants have sex organs with an outer layer of nonreproductive cells that can prevent desiccation of gametes - the flower. The developing diploid embryo is protected from drying out by providing it with water and nutrients within the female reproductive structure - the fruit. Growth in plants can be indeterminate. All plants have a two-generation life cycle known as alternation of generations. This means that a plant exists in 2 forms: the haploid generation is the gametophyte that produces gametes; and the diploid generation is the sporophyte that produces spores by meiosis. Spores are haploid structures that develop or mature into the gametophyte plant. Usually, lower plants spend their life longer in haploid (1N) generation.

In addition to a cell membrane, plant cells are surrounded by a cell wall that varies in thickness depending on the function of the cell. All plant cells have a primary cell wall, having as its main constituent cellulose molecules united into threadlike microfibrils. Several microfibrils, in turn, are found in fibrils, and within the wall there are layers of fibrils lying at right angles to one another for added strength. Some cells in woody plants have a secondary cell wall that forms inside the primary cell wall. Secondary cells contain lignin (major component of wood), a substance that makes secondary cell walls even stronger than primary cell walls. Plant support cells have secondary walls; the cell dies and the strong wall remains as supporting material. The plasmodesmata in Figure 06 is a very fine thread of cytoplasm that passes through openings in the walls of adjacent cells and forms a living bridge between them.

General Structures of Plants

The Root System:

- It is usually underground.
- It anchors the plant in the soil.
- The root absorbs then conducts water and minerals to the stem.
- It is a food storage.

The Shoot System:

- It is usually above ground to elevates the plant from the soil.
- The shoot system includes the stem, the leaves, and the reproductive organs.

It provides many functions including:

- photosynthesis.
- reproduction and dispersal.
- nutrients and water conduction.

Algae and Classification of Plants

The term algae is used for aquatic unicelluar organisms that photosynthesize as do terrestrial plants. All algae contain green chlorophyll, but they also can contain other pigments that mask the color of the chlorophyll. The chlorophylls selecting different part of the spectrum are separated into type a, b, c, and d. Thus there are green, golden brown, brown, and red algae. Green algae can be single-celled, colonial, filamentous, and multicellular. It is believed that the green algae are ancestral to the first plants because both of these groups possess chlorophylls a and b, both store reserve food as starch, and both have cell walls that contain cellulose. Seaweeds such as Ulva are multicellular algae carrying chlorophyll. They are anchored firmly to the rock by holdfasts. Their appearance give a false impression that they are plants having root, stem, and leaf.

While a few of the algae such as some species of Fucus follow the diplontic lifecycle, most of them such as the green algae Chlamydomonas spend most of their life in the haploid generation. Usually, this protist practices asexual reproduction, and the adult divides to give zoospores that resemble the parent cell. During sexual reproduction, gametes of two different strains come into contact and join to form a zygote. A heavy wall forms around the zygote, and it becomes a zygospore. The zygospore is able to survive until conditions are favourable for germination and subsequent production of 4 zoospores by meiosis. The gametes that they look exactly alike. Sexual reproduction aids the process of evolution because it offers means to produce variations in addition to mutations. Note that embryo is absent in this kind of lifecycle Bryophytes (Moss, Liverwort).

The bryophytes include liverworts and mosses. Most species of liverworts are "leafy" and look somewhat like mosses, but close examination shows that the body of a liverwort has distinct top and bottom surfaces, with numerous rhizoids (rootlike hairs) projecting into the soil. In contrast, a moss has a stemlike structure with radially arranged, leaflike structures. Rhizoids anchor the plant and absorb minerals and water from the soil. Because bryophytes do not have vascular tissue, they lack true roots, stems, and leaves. Instead, they have rhizoids, stemlike and leaflike structures . In mosses, the gametophyte is dominant - it is longer lasting. In some mosses, there are separate male and female gametophytes). At the tip of a male gametophyte are antheridia, in which swimming sperms are produced. After rain or heavy dew, the sperm swim to the tip of a female gametophyte, where eggs have been produced within the archegonia.

Antheridia and archegonia are both multicellular structures, and each has an outer layer of jacket cells that protects the enclosed gametes from drying out. After an egg is fertilized, the developing sporophyte is retained within the archegonium as an embryo. The sporophyte, which is dependent on the gametophyte, consists of a foot that grows down into the gametophyte tissue, a stalk (seta), and an upper capsule, or sporangium, where meiosis occurs and where haploid spores are produced. In some species of mosses, a hoodlike covering is carried upward by the growing sporophyte. When this covering and the capsule lid falloff, the spores are mature and ready to escape. The rings of "teeth" projected inward from the edge of the capsule allows spores to be released only at times when the weather is dry (when they are most likely to be dispersed by wind). When a spore lands on an appropriate site, it germinates. The single row of cells that first appears branches, giving an algalike sturcture called a protonoma. After about three days of favorable growing conditions, new moss plants appear at intervals along the protonema. Each of these consists of the rootlike rhizoids and the upright shoots of a moss gametophyte. The gametophytes produce gametes, and the moss life cycle begins again.

Sperms are released when the antheridium ruptures, thus allowing them to swim freely in a water film toward the archegonium. The zygote is the first cell of the new sporophyte just after fertilization. The zygote divides by mitosis into a multicellular embryo within the archegonium). This is the crucial step that separates plants from algae. The embryo then inserts an absorbing organ called the foot into the female stem tip. The other end of the embryo grows up and above the female stem to form a stalk (seta) and sporangium (capsule) anchored in the old archegonium. Early in its development the sporophyte is typically green, but by the time it is mature it is usually non-photosynthetic and dependent on the gametophyte for water and nutrients. Within the sporangium special cells called sporocytes divide by meiosis to produce thick-walled haploid spores. In the more advanced plant species, the embryo is enclosed within the seed.

PTERIDOPHYTES (FERN, CLUB MOSS, HORSETAIL)

Vascular plants (also called tracheophytes) are believed to have evolved sometime during the late Silurian Period. The primitive vascular plants include the whisk ferns (Psilopsid), the club mosses, and the horsetails . The whisk ferns is of particular interest because they may be the most primitive. It bears considerable resemblance to the extinct rhyniophytes. Its sporophyte consists of stems with scalelike structures but no leaves. There is a horizontal stem (lacking roots), from which rhizoids grow, and there are green, photosynthetic, upright branches with tiny, scalelike structures that grow upward. Sporangia are located on the branches. The gametophyte is separate from and smaller than the sporophyte; it also lacks vascular tissue. In general, the tracheophytes have two types of vascular tissue. Xylem conducts water and minerals up from the soil, and phloem transports organic nutrients from one part of the body to another. Because they have vascular tissue, the specialized body parts of tracheophytes can be called properly roots, stems and leaves.

The life cycle of a common fern of the temperate zone. Young fronds grow in a curled-up form called fiddleheads, which unroll as they grow. The fronds often are subdivided into a large number of leaflets. The sporophyte fern plant represents the dominant generation. Sporangia develop in clusters called sori, which are protected by a covering, the indusium (not shown). Within the sporangia, meiosis occurs and spores and produced. The gametophyte is a tiny (1-2 cm), heart-shaped structure called a prothallus. The antheridia and archegonia develop on the under side of a prothallus. Fertilization takes place when moisture is present because the spiral-shaped sperm must swim from the antheridia to the archegonia. The resulting zygote soon develops into a sporophyte embryo consisting of a foot, a root, a stem, and a leaf. The root grows down into the soil, and the frond grows upward through the prothallus notch. As the sporophyte matures, the prothallus shrivels and disappears. Since the gametophyte lacks vascular tissue, and the swimming sperm relies on moisture to approach the egg, ferns are likely to be found in habitats that are at least seasonally moist. Once established, the sporophyte of some ferns can spread by vegetative reproduction into drier areas because this generation has vascular tissue.

The vascular structures in the ferns are primitive in comparison to the more advanced plant species. They have the rhizome, which can be compared to the stem of a flowering plant. In many cases the rhizome can be inconspicuous or even entirely underground. Rhizomes of tree ferns on the other hand may be 60 cm in diameter and up to 12 meters tall. The fronds (leaves) arise from the upper side or in one or more rows laterally on each side from the rhyzome. They are composed of two main structures: the stipe (stalk) and the blade (the leafy outcroppings). Roots are formed from the rhizomes or sometimes from the stipe. The roots usually do not divide once they grow from the rhizome. Tree fern roots grow down from the crown and help thicken and strengthen the trunk. The roots anchor the plant to the ground and absorb water and minerals.

In the more advanced plant species, the outermost tissue of the stem is the epidermis. The stem has distinctive vascular bundles, where xylem and phloem are found. In each bundle, xylem is typically found toward the inside and phloem is toward the outside. In the dicot stem, the bundles are arranged in a distinct ring that separates the cortex from the central pith The cortex is sometimes green and carries on photosynthesis, and the pith may function as a storage site for the products of photosynthesis. In the monocot stem, the vascular bundles are scattered throughout the stem, and there is no well-defined pith. Secondary growth of stems is seen primarily in woody plants, such as trees that live for many years. Almost all trees are dicots. Primary growth in woody plants occurs for a short distance beneath the apical meristem. Secondary growth occurs in the vascular and cork cambia). Vascular cambium begins as meristematic cells between the xylem and the phloem of each vascular bundle. Then these cells join to form a ring of meristematic tissue adding to the girth of the stem. Cork cambium is located beneath the epidermis. It produces tissue that disrupts and replaces the epidermis with cork cells, which are impregnated with suberin (a waterproof substance). Dead cork allows gas exchange in pockets of loosely arranged cells, called lenticels. A woody stem has three distinct areas: the bark (containing cork, cork cambium, cortex, and phloem), the wood, and the pith. In large trees, only the more recently formed layer of xylem, the sapwood, functions in water transport. The older inner part, called the heartwood, becomes plugged with deposits, such as resins, gums, and other substances.

A longitudinal section of a root. At the bottom is an area of cells called the root cap, a thimble-shaped mass of parenchymal cells (relatively unspecialized cells) that is a protective covering for the root tip, and the cells in the next region – the region of cell division. Cells in the root cap have to be replaced constantly because they are ground off as the root pushes through abrasive soil particles. The next area - the zone of cell division - is the area where new cells are continually being formed through repeated cell divisions. These cells are thin-walled and easily ruptured by soil particles were it not for the root cap's protection. Next is the zone of cell elongation. Here the cells take up large amounts of water and increase in volume. The increase in cell volume of these cells is primarily responsible for pushing the root through the soil. The next zone is the zone of cell maturation and differentiation. The fully elongated cells in these zones matured and began differentiating into various tissues such as the xylem, phloem, pith, cortex, and others. This zone, the zone of maturation and differentiation begins where the root hairs first become evident. Branch roots have formed beyond these zones.

The absorbed water and minerals pass through the cortex, a tissue composed of parenchymal cells. The water and minerals are forced by a strip of waxy material (the Casparian strip) in the endodermis to move one way into the vascular cylinder. Within the vascular cylinder, water and minerals are transported upward by way of the xylem and the products of photo-synthesis most often are transported downward by way of the phloem for storage in the cortex. Lying between the endodermis and the vascular tissue is the pericycle, composed of parenchymal cells, that retains the ability to undergo cell division and on occasion produces branch roots. The pericycle alos contributes to the formation of vascular cambium, which is meristematic tissue lying between xylem and phloem that is capable of producing new vascular tissue. Monocot roots often have pith, which is centrally located ground tissue. In a monocot root, pith is surrounded by a ring of alternating xylem and phloem bundles. They also have pericycle, endodermis, cortex, and epidermis.

GYMNOSPERMS (CYCED, GINKO, CONIFERS)

The gymnosperms produce naked seeds; that is, the seeds are not enclosed by fruit. There are four divisions of gymnosperms. Cycads are cone-bearing, palmlike plants found today mainly in tropical and subtropical regions. Only one species of ginkgo, the maidenhair tree, survives today. The gnetophyta has only three genera left. The largest group of gymnosperms is the cone-bearing conifers, which include pine, cedar, spruce, fir, and redwood trees. These trees have needlelike leaves that are well adapted to not only hot summers but also cold winters and high winds. Most gymnosperms are evergreen trees.

The sporophyte is dominant in the pine life cycle. Typically, the male pine cones are quite small and develop near the tips of lower branches. Each scale of the male cone has two or more microsporangia on the underside. Inside the microsporangia are microspore mother cells that undergo meiosis and develop into mature pollen grain (with two lobular wings), which is a sperm-bearing male gemetophyte. The female pine cones are larger and located near the top of the tree. Each scale of the female cone has two ovules that lie on the upper surface. Within the ovule, a megaspore mother cell undergoes meiosis and develops into mature female gametophyte, which has 2-6 archegonia, each containing a single, large egg lying near the ovule opening. During pollination, pollen grains are transferred from the male cone to the female cone. Once enclosed within the female cone, the pollen grain develops a pollen tube that slowly grows toward the ovule. The pollen tube discharges two nonflagellated sperms. Only one of the sperms fertilizes an egg in the ovule 15 months after pollination. After fertilization, the ovule matures and becomes the seed composed of the embryo, its stored food, and a seed coat. Finally, in the third season, the female cone, by now woody and hard, opens to release its seeds, whose wings are formed from a thin, membranous layer of the cone scale. When a seed germinates, the sporophyte embryo develops into a new pine tree, and the cycle is complete.

The outermost layer of the conifer seed is the seed coat. It originates from the mothe tree and is diploid. The seed coat has three layers: the outer layer; the thicker, tough stony or middle layer; and the inner layer. Some species of conifers have resin vesicles in the middle or outer layers of the seed coat. These resin vesicles may play a role in seed coat dormancy, protecting dehydration, and deterring seed herbivory. Immediately inside the cell wall is the nucellus, a papery layer surrounding the megaspore cell wall. Inside the megaspore cell wall is the megagametophyte, the haploid nutritional tissue found in gymnosperm seeds. Early in the development of cones, megagametophytes produce egg cells which are fertilized by male gemetes to produce zygotes. Zygotes develop into embryos. The megagametophyte then plays its second functional role, surrounding the embryo, protecting and nourishing it. The megagametophyte in Douglas-fir is 60% lipids, 16% proteins, and 2% sugars, making it a high-energy and nutrient tissue both for the embryos it contains and for a plethora of seed predators including small mammals, birds, and insects. The embryo is found in the corrosion cavity, a pit in the centre of the megagametophyte that is fully filled by the embryo in mature seeds. It consists of the cotyledons (first leaf), shoot apical meristem, root apical meristem, root cap and suspensor. The cotyledons and shoot apical meristem point towards the wider end of the seed; whıle the radicale (embryonic root) and suspensor are at the more pointed end. The suspensor is found at the base of the root cap and plays a role early in embryo development by pushing the embryo into the megagametophyte.

ANGIOSPERMS (OAK, MAPLE, BASIL, ...)

Angiosperms are the flowering plants with seeds enclosed by fruit. All hardwood trees (broad-leaved trees, e.g., oak, ...), including all the deciduous trees (trees that shed their leaves in the fall, e.g., maple, ...) of the temperate zone and the broad-leave evergreen trees (e.g., boxwood, ...) of the tropical zone, are angiosperms, although sometimes the flowers are inconspicuous. All herbaceous (nonwoody, nonpersistent) plants (e.g., basil, ...) common to our everyday experience, such as grasses and most garden plants are flowering plants . Angiosperms are adapted to every type of habitat, including water (e.g., water lilies, ...). Angiosperms have well-developed vascular and supporting tissues. Their xylem tissue contains vessel elements beside the tracheids. Thus the woody angiosperms are considered as hardwood trees, whereas the gymnosperms are softwood trees.

In angiosperms, the reproductive structures are located in the flower. The flower attracts insects and birds that aid in pollination, and it produces seeds enclosed by fruit. There are many different types of fruits, some of which are fleshy (e.g., Apple, tomato, peach, ...) and some of which are dry (e.g., pea enclosed by pod, nut, grain, ...). They all provide protection for the seeds. Within a flower, there is a diploid megaspore mother cell in each ovule of the ovary. The mother cell undergoes meiosis, producing one functional megaspore, whose nucleus divides mitotically until there are eight haploid nuclei. This is the female gemetophyte, which sometimes is called the embryo sac. At one end of the embryo sac there the three cells, one of which is the egg cell. Male gametophytes are produced in the stamens. An anther contains four pollen sacs with many microspore mother cells, each of which undergoes meiosis to four microspores. After a mitotic division, each misrospore has two cells, one of which later divides again to give two sperm. Pollination, which is simply the transfer of pollen from the anther to the stigma, is brought about by wind or with the assistance of a particular pollinator. The plant uses the pollinator to ensure cross-pollination, and the pollinator uses the plant as a source of food in the form of nectar. When a pollen grain lands on a stigma of the same species, it germinates, forming a pollen tube. The pollen tube grows as it passes between the cells of the stigma and the style to reach the female gemetophyte.

Double fertilization takes place to produce seeds and fruits. One sperm nucleus from the pollen tube unites with the egg nucleus, forming a zygote, and the other sperm nucleus unites with the polar nuclei, forming a triploid (3N) endosperm nucleus. The endosperm nucleus divides, forming the endosperm, which is a nutrient material for the developing embryo and sometimes for the young seedling as well. The zygote develops into an embryo. The outer layers (integuments) of the ovule harden and become the seed coat. A seed is a structure formed by the maturation of the ovule; it contains a sporophyte embryo plus stored food. The ovary and sometimes other floral parts develop into the fruit. A fruit is a mature ovary that usually contains seeds. Therefore, angiosperms are said to have covered seeds.

6

Fern Identification

1. Silvery Spleenwort - *Athyrium thelypteroides*

The silvery spleenwort is dull green with a hairy stem. It prefers rich, moist woodland ravines.

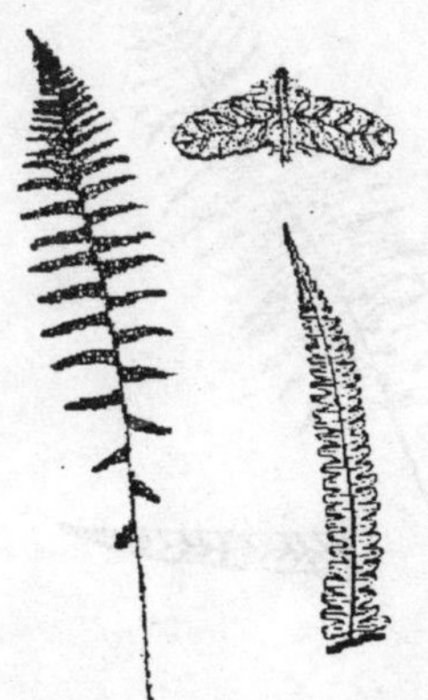

2. Marginal Woodfern - *Dryopteris marginalis*

This bluish-green, somewhat leathery evergreen is often found hanging from rocks and ledges on north-facing slopes. It is common in our county.

3. Narrow-Leaved Spleenwort - *Athyrium pycnocarpon*

A fairly slender fern about waist high, with smooth slender leaflets and stem, the narrow-leaved spleenwort grows in deep, moist, rich woodlands.

4. Netted Chain Fern - *Woodwardia areolata*

A glossy green little fern with margins of leaflets fine-toothed, the netted chain fern can be found in the shade of swamps and wet woods

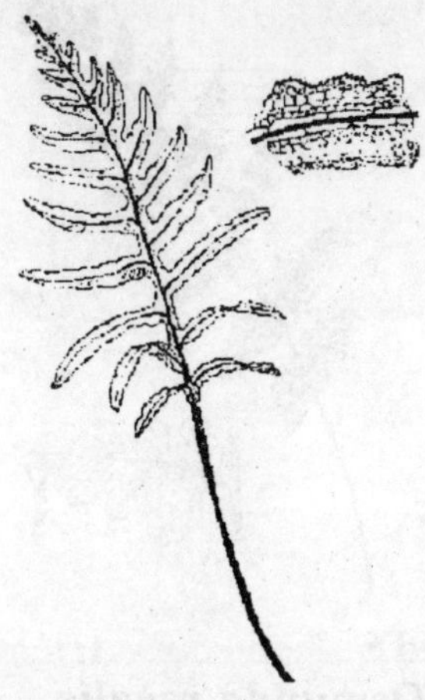

5. **Hayscented Fern - *Dennstaedtia punctilobula***

A yellowish-green brittle fern, this plant is often found growing in colonies in deep ravines and cliffs along water ways. It also hangs from cliffs.

6. **Fragile Fern - *Cystopteris fragilis***

This bright green small fern, usually arching from a moss bank, often disappears during a dry spell. Its stem is brittle to the touch.

7. Royal Fern - *Osmunda regalis*

The pale green leaves of this fern grow up to 2 meters in length, with its roots forming large basal cushions. It is found in wet woods and along woodland streams, often growing in shallow water.

8. Cinnamon Fern - *Osmunda cinnamomea*

A large fern standing about shoulder high, the cinnamon fern is seen growing in wet swampy woods and along stream edges.

9. New York Fern - *Thelypteris noveboracensis*

This yellow-green fern grows in colonies about knee high in sunlit woodlands. Its leaflets are gradually reduced down the stem.

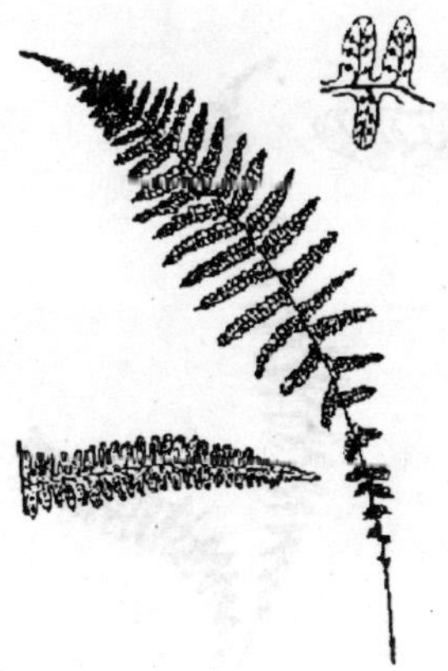

10. Walking Fern - *Camptosorus rhizophyllus*

The walking fern is a little evergreen fern with leaves flat on the ground. The tips of its leaves form new little fern plants. It is found only on wet rocky areas.

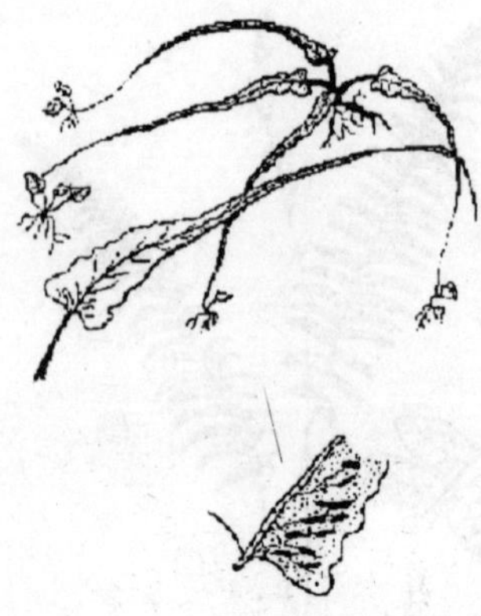

11. Ebony Spleenwort - *Asplenium platyneuron*

One of our commonest ferns of dry woodland and woodland edges, the ebony spleenwort's alternate leaflets and shining brown stems are conspicuous.

12. Maidenhair Spleenwort - *Asplenium trichomanes*

This is a small fern inhabiting moist shaded crevices in rock outcroppings.

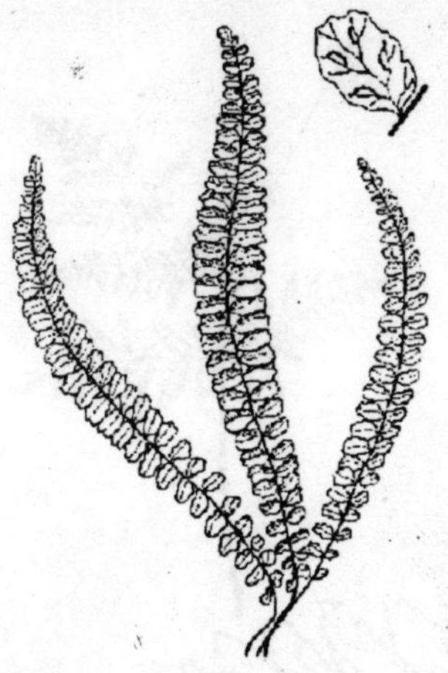

13. Maidenhair Fern - *Adiantum pedatum*

The maidenhair fern has delicate horseshoe-like fronds, a shining black stalk, and is in shape unlike any of our other local ferns. It grows in rich shaded soil, often in ravines.

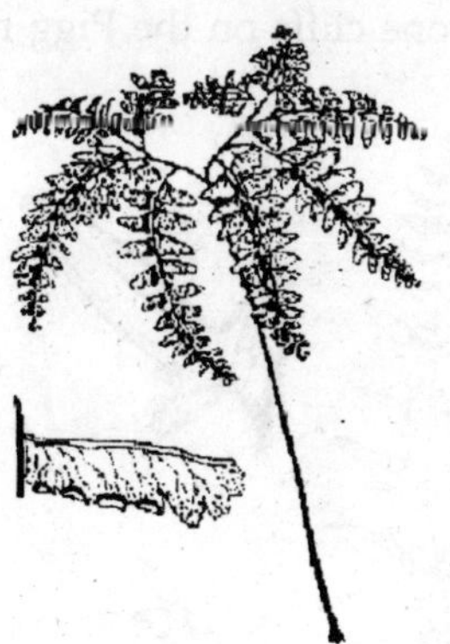

14. Rattlesnake Fern - *Botrychium virginianum*

This common fern is a bright green, lacy-cut early-appearing fern of the woodlands. Its spore shoot arises from upper axil of the leaf.

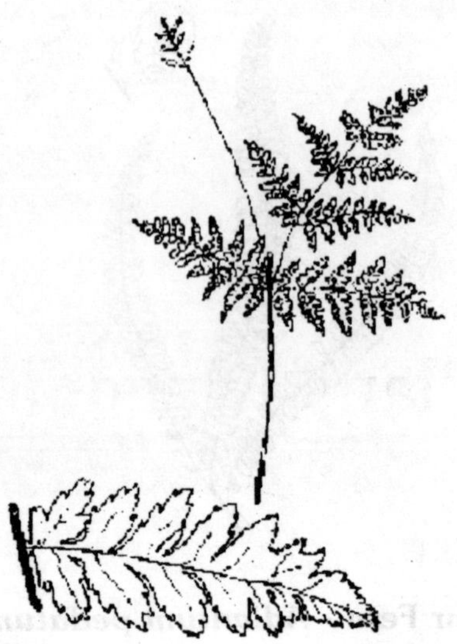

15. Purple-Stemmed Cliffbrake - *Pellaea atropurpurea*

This evergreen is rare in our country, only found growing from crevices in limestone cliffs on the Pigg river.

7

Plant Kingdom

CHARACTERISTICS

- Eukaryotic
- Photosynthetic
- Multicellular
- Sexually reproducing
- Life History involves an alternation of a haploid phase (Gametophyte) with a diploid phase (Sporophyte)

  Classi f icat ion Artificially grouped into Nonvascular or Vascular Plants Non Vascular Plants – "Bryophytes"
- Disperse by spores
- Require H_2O for sexual fertilization
- Small
- No vascular tissue
- Grow in clumps or masses
- Many capable of prolonged dehydration
- Most lack cuticle (protective surface layer)
- Three Divisions of Bryophytes

  Three Phyla

Bryophyta – Mosses

Hepatophyta – Liverworts

Anthocerophyta – Hornworts

Vascular Plants

- Vascular tissue
- Cutin or suberin on aerial parts
- Body an axis (stern)
- Large surface/volume ratio common
- Much survival value for terrestrial habitat with water deficit – increase in size by conducting tissue – especially xylem
- Many subdivision and classes
- Each textbook varies, but – patterns are visible as increase in complexity of body structure
- Vascularization (Microphylls vs megaphylls in leaves)
- Dispersal {spore (Single cell) vs seed}
- Sperm transport H_2Ovs pollen grain)

Lower (spore–dispersing) Vascular Plants

Mostly predominant in fossil eras

Important in coal–formation

Fossil Phyla

Rhyniophyta

Zosterophyl lophyta

Tr imerophytophyta

Extant Phyla

Lycophyta (Lycopodiophyta) Ground Pines, Club Mosses and Quillwort

Three Extant Classes (Orders) (Families)

Lycopodiae (Lycopodiaes) (Lycopodiaceae)

Selaginellae (Selaginellales) (Selaginellaceae)

Isoetae (Isoetales) (Isoetaceae)

Pter idophyta

Ferns (Pterophyta)

Five Orders

Ophioglossales – Eusporangiate

Marattiales – Eusporangiate

Filicales – Homosporous and Leptosporangiate

Salviniales – Heterosporous and Leptosporangiate

Marsileales – Heterosporous and Leptosporangiate

Horsetails or Equisetum – One Order

Equisetales (Sphenophyta)

Whisk Ferns – One Order

Psilotales (Psilotophyta)

Higher (seed-dispersing) Vascular Plants

- Disperse by multicellular seeds
- Embryo head–start in seed package
- Transport sperm in pollen grain
- Freedom from H2O for fertilization

Two Broad Groups plus the Fossils

Fossil Phyla

Progymnospermophyta

Pteridospermales – Fossil Seed Ferns

Cordaitales – Primitive Conifer-like

Bennettitales – Fossil Cycadeoides

Extant Phyla

Gymnosperms – Seed not protected by a fruit

Four Phyla

Cycadophyta: Cycads

Coniferophyta: Conifers (e.g., Pine, Spruce, Fir, Hemlock, Yew)

Ginkgophyta: Ginkgo

Gnetophyta: Gnetum, Ephedra, Welwitschia

Angiosperms (Flowering plants) – Seed protected by a fruit (the ovary)

One Phylum Anthophyta – Flowering Plants

Two Classes comprise 97% of Angiosperms

Eudicotyledones (Dicots)

Monocotyledones (Monocots)

Additional Groups comprise 3% of the more primitive Angiosperms

Magnoliidae, including several orders

Nymphaeales – Water Lilies

Illiciales – Star Anise

Amborel lales

Archaefructales – Earliest Anthophyte Fossil

Characteristics of the Plant Groups

Non-Vascular Plants

The Bryophytes (Mosses, Hornworts and Liverworts)

- Multicellular
- Non-vascular
- Haploid generation dominant, assimilative phase
- Mostly terrestrial, of moist habitats
- Water necessary for sexual reproduction
- Flagellated sperm cells
- Archegonium - female sex organ

- Antheridium - male sex organ
- All small!

Phylum – Hepatophyta (Liverworts)

- Two groups

Thallose

Leafy (resemble mosses, but lack a midrib on the "leaves", and have dorsalventral symmetry

- Dorsal-ventral symmetry
- Vegetative reproduction by

Fragmentation

Gemmae

- Sporangium simple

Phylum – Bryophyta (Mosses)

- Radially symmetrical
- Gametophyte phyllodes ("leaves") have a midrib (costa)
- Sporangium a capsule elevated by a stalk or seta above the gametophyte

Phylum – Anthocerophyta (Hornworts)

- Rounded thallose-like gametophyte
- Vegetative reproduction by
- Fragmentation
- Sporophyte "horn-shaped" growing from a basal sheath beneath the surface of the gametophyte thallus. The sporophyte continues to grow from a basal meristem, producing spores clustered around a central stalk. The sporophyte tip splits releasing spores
- Less common than liverworts or mosses

Vascular Plants

Phylum – Lycophyta (Lycopodiophyta) (Club mosses)

- True roots, stems and leaves (Microphylls)
- Sporangia borne in a strobilus
- May be Homosporous (One type of sporangium) or Heterosporous (Produce megasporangia and microsporangia)
- Strobili may be axillary or terminal
- Examples

Lycopodium

Selaginella

Isoetes

Phylum - Pterophyta (Ferns, Whisk Ferns and Horsetails)

Ferns

- Sporophyte generation dominant
- Both gametophyte and sporophyte independent
- Underground rhizome typical (Stem)
- Fertile and sterile fronds (Leaves {Megaphylls})
- Spores produced in sporangia located in a sorus which usually has a protective indusium
- Sori located on underside of fertile fronds
- Several Orders

Ophioglossales – Eusporangiate

Marattiales – Eusporangiate

Filicales – Homosporous and Leptosporangiate

Salviniales – Heterosporous and Leptosporangiate

Marsileales – Heterosporous and Leptosporangiate

Order – Equisetales (Sphenophyta) (Horsetails and Scouring rushes)

- Stems silica impregnated
- Branches whorled

- Stems jointed
- Leaves microphylls and often non-photosynthetic
- Homosporous (Archegonia and Antheridia on same plant)
- Spores have elaters for dispersal
- One Genus

Equisetum

Order – Psilotales (Psilotophyta)

- No true roots or leaves
- Phyllodes present
- Sporangia borne in axils, in clusters
- Determined to be a degenerate group related to ferns
- Examples

Psilotum

Tmesipteris

Phylum – Progymnospermophyta

Order – Pteridospermales (Pteridospermophyta) (Seed Ferns)

- All fossils
- Important as the probable progenitors of today's seed plants

Phylum - Cycadophyta (Cycads)

- 9 genera, 100 species
- Tropic and subtropical
- Vegetative characteristics:
- Stem unbranched, short or columnar
- Terminal crown of long, leathery, compound leaves (Palm-like)
- Less than 6 feet tall (mostly)
- Reproduction

- All Dioecious (Male and Female Strobili on separate plants)
- Pollen sacs on scalelike microsporophylls in compact cones (Microsporangia)
- Megasporophylls also in cones (some up to 3 ft in length)
- Pollination by wind, but male gamete motile
- Largest motile gametes with up to 20,000 flagella
- Examples

Cycas revoluta (Sago palm)

Zamia

Phylum - Ginkgophyta (Ginkgo or Maidenhair tree)

- One living species Ginkgo biloba
- Deciduous tree with fan-shaped leaves
- Reproduction
- Sexual cycle like cycads
- Male strobilus in short pendant paired microsporophylls
- Female Ovules develop into yellowish, cherry-like seeds
- Seed coat decomposes at maturity for dispersal

Phylum - Gnetophyta

- Do not produce flowers
- Seeds not protected by a fruit (ovary)
- Xylem contains vessels
- Oldest fossils just 50 million years BP
- Three representatives (All distinctive)

Gnetum

Ephedra

Welwitschia

Phylum - Coniferophyta (Conifers)

- Dominant vegetation of the Taiga regions

- Major source of paper and lumber
- Strobili modified into cones
- Cone scales are modified sporophylls
- Xylem contains tracheids but no vessels
- Reproduction
- Male and female gametophytes produced in separate cones
- Female cones contain Megasporangia
- Each megasporangium produces an ovule with an integument
- Meiosis produces a megaspore, retained within the cone that develops into the female gametophyte with archegonium containing an egg
- Male cones (Microsporangia in strobilus) contain Microspores that develop into the pollen grains (male gametophytes)
- Wind pollination – Water unnecessary for fertilization
- Seeds unprotected by sterile tissue
- Examples

Pinus Pseudotsuga Metasequoia

Abies Tsuga Taxus

Picea Thuja Juniperus

Phylum - Anthophyta (Flowering Plants)

- Reproduction
- Reproductive organs within a flower rather than a cone
- Ovule embedded in sporophyte tissue (ovary)
- Gametophytes greatly reduced
- Fertilization "double"
- One sperm with the egg:
- One sperm with polar nuclei to form a nutritive endosperm

- Seeds enclosed in a fruit (ovary)

  Two Major Classes (Distinguished anatomically)

Monocotyledones (Monocots)	Eudicotyledones (Dicots)
Flower parts in 3's	Flower parts in 5's (4's)
Leaves with parallel veins	Leaf venation palmate or pinnate
No true cambium	Cambium usually present
One cotyledon	Two cotyledons
Scattered vascular bundles	Vascular bundles in a ring (cylinder)
Sheathing leaf bases	Leaves usually have a petiole

Additional Groups of more primitive Angiosperms

Magnoliidae, including several orders

Nymphaeales – Water Lilies

Illiciales – Star Anise

Amborellales

Archaefructales – Earliest Anthophyte Fossil.

8

Frond and Cycad

INTRODUCTION

A frond is a large leaf with many divisions to it, and the term is typically used for the leaves of palms, ferns or cycads.

A frond is the leaf- like structure of a fern or alga. The term is colloquially applied to the leaves of palms, cycads, and plants with pinnately compound leaves. A significant difference is that, unlike the leaves of the latter, fern fronds bear the reproductive structures (spore-bearing structures) of the sporophyte plant. Because many ferns grow fronds that are held more vertical than horizontal, the "upper" and "lower" surfaces of a frond are more correctly referred to as the adaxial and abaxial surfaces, respectively.

A fern frond consists of a stipe, the stem supporting the blade, and the blade consists of both a laminar (flattened) photosynthetic tissue and a rachis—that portion of the stem to which the laminar tissue is attached. The blades of fern fronds may vary from being simple (undivided) to being highly dissected, even "lace-like". If the leaf tissue is undissected, or the dissections do not reach to the *rachis*, the frond may be described as lobed or *pinnatifid*. Otherwise, the blade is compound and each large division of the laminar tissue arising from the rachis is called a pinna (pl., pinnae). The main vein or mid-rib of a pinna is known as a costa (pl., costae). Pinnae may be arranged along the *rachis* either directly opposite one another or alternating up the stem. The arrangement may change from the base of a blade to the tip, as in the example of *Blechnum* shown below (from base to tip: pinnae opposite to alternate, and pinnatisect to pinnatifid).

Many ferns have *pinnae* that are divided two or more times, and the level of division of the fronds is termed *pinnate* (or 1-pinnate), or *twice-pinnate* (2-pinnate), or the like. Each secondary division (division of a pinna) is termed a pinnule, and its mid-vein, a costule. A few species of ferns with divided fronds are not pinnate, but are palmate or bifurcate.

On some or all mature blades (usually on the *abaxial* surface) occur sporangia, which bear the spores. The sporangia are clustered in a sorus (pl., sori) or "fruit dot". Associated with each sporus in many species is a membranous structure called an indusium: an outgrowth of the blade surface that may partly cover the sporangial cluster. Fronds also may bear hairs or scales, glands, and, in some species, bulblets for vegetative reproduction.

Each frond arises from the stem or rhizome, which in most species is concealed in the ground or creeps along the ground (or branch or rock) surface. Growth of a fern frond differs from that of a leaf of a flowering plant. The fern frond unrolls from a tightly-coiled structure called a "fiddle-head" (see circinate vernation).

Some fern species feature frond dimorphism, in which fertile and sterile fronds differ in appearance and structure.

CYCAD

Cycads are a group of seed plants characterized by a large crown of compound leaves and a stout trunk. They are evergreen, gymnospermous, dioecious plants having large pinnately compound leaves. They are frequently confused with and mistaken for palms or ferns, but are related to neither, belonging to the division Cycadophyta.

Cycads are found across much of the subtropical and tropical parts of the world. They are found in South and Central America (where the greatest diversity occurs), Mexico, the

Antilles, south-eastern United States of America, Australia, Melanesia, Micronesia, Japan, China, Southeast Asia, India, Sri Lanka, Madagascar, and southern and tropical Africa, where at least 65 species occur. Some are renowned for survival in harsh semi-desert climates, and can grow in sand or even on rock. They are able to grow in full sun or shade, and some are salt tolerant. Though they are a minor component of the plant kingdom today, during the Jurassic period they were extremely common.

They have very specialized pollinators and have been reported to fix nitrogen in association with a cyanobacterium living in the roots. This blue-green algae produces a neurotoxin called BMAA that is found in the seeds of cycads.

Origins

The cycad fossil record dates to the Early Permian, 280 mya. There is controversy over older cycad fossils that date to the late Carboniferous period, 300–325 mya. One of the first colonizers of terrestrial habitats, this clade probably diversified extensively within its first few million years, although the extent to which it radiated is unknown because relatively few fossil specimens have been found. The regions to which cycads are restricted probably indicate their former distribution on the supercontinents Laurasia and Gondwana.

The family Stangeriaceae (named for Dr. William Stanger, 1812–1854), consisting of only three extant species, is thought to be of Gondwanan origin as fossils have been found in Lower Cretaceous deposits in Argentina, dating to 70–135 mya. Zamiaceae is more diverse, with a fossil record extending from the Middle Triassic to the Eocene (54–200 mya) in North and South America, Europe, Australia, and Antarctica, implying that the family was present before the break-up of Pangea. Cycadaceae is thought to be an early offshoot from other cycads, with fossils from Eocene deposits (38–54 mya) in Japan and China, indicating that this family originated in Laurasia. *Cycas* is the only genus in the family and contains 99 species, the most of any cycad genus. Molecular data has recently shown that *Cycas* species in Australasia and the east coast of Africa are recent arrivals, suggesting that adaptive radiation may have occurred. The current distribution of cycads may be due to radiations from a few ancestral types sequestered on Laurasia and Gondwana, or could be explained by genetic drift following the separation of already evolved genera. Both explanations account for the strict endemism across present continental lines.

TAXONOMY

There are about 305 described species, in 10–12 genera and 2–3 families of cycads (depending on taxonomic viewpoint). The classification below, proposed by Dennis Stevenson in 1990, is based upon a hierarchical structure based on cladistic analyses of morphological, anatomical, karyological, physiological and phytochemical data.

The number of species in the clade is low compared to the number of species in most other plant phyla. However, paleobotanical and molecular research indicates that diversity was higher in the history of the phylum. Fossil evidence shows that structural diversity in Mesozoic cycad pollen "considerably exceeds that seen in surviving genera today". The impacts of extinction on diversity are highlighted below. The disparity in molecular sequences is very high between the three main lineages of cycads, implying that genetic diversity in the clade was once high, but this fact has led to major disagreements about the divisions within the Cycadales.

The number of described cycad species has doubled in the past 25 years, mostly due to improved sampling and further exploration. Experts assume there may still be about 100 undescribed species, based on the rate of discovery. These are likely to be in Asia and South America where areas of endemism are highest. Diversity hotspots also occur in Australia, South Africa, Mexico, China and Vietnam, which together account for more than 70% of the world's cycad species. The taxonomy of the Cycadophyta is, however, now stabilizing.

Cycad systematists reject the biological species concept, as clearly defined cycad species can interbreed and produce fertile offspring; this character is thus not disproportionately weighted when determining species barriers. The phenetic species concept, which states that a species is defined based on overall similarities with other individuals of the same species combined with a significant gap in variation with other species, is also rejected. Most cycad taxonomists agree on a modified version of the evolutionary species concept, termed the 'morphogeographic' species concept, which recognises the combined effects of geographical isolation and morphological disparity. Thus the presence of large geographical gaps in cycad distribution has greatly affected the way cycads are classified:

Suborder: Cycadineae

Family: Cycadaceae

Subfamily: Cycadoideae

Cycas. About 105 species in the Old World from Africa east to southern Japan, Australia and the western Pacific Ocean islands; type: *C. circinalis* L.; see also *C. pruinosa* and *C. revoluta*

Suborder: Zamiineae

Family: Stangeriaceae

Subfamily: Stangerioideae

Stangeria: One species in southern Africa; type: *S. eriopus* (Kunze) Baillon

Subfamily: Bowenioideae

Bowenia: Two species in Queensland, Australia; type: *B. spectabilis* Hook. ex Hook. f.

Family: Zamiaceae

Subfamily: Encephalartoideae

Tribe Diooeae

Dioon. 13 species in Mexico and Central America; type: *D. edule* Lindley

Tribe: Encephalarteae

Subtribe: Encephalartinae

Encephalartos. About 66 species in southeast Africa; type: *E. friderici-guilielmi* Lehmann, *E. transvenosus (Modjadji cycad)*

Subtribe: Macrozamiinae

Macrozamia. About 41 species in Australia; type: *M. riedlei* (Fischer ex Gaudichaud) C.A. Gardner

Lepidozamia. Two species in eastern Australia; type: *L. peroffskyana* Regel

Subfamily: Zamioideae

Tribe: Ceratozamieae

Ceratozamia. 26 species in southern Mexico and Central America; type: *C. mexicana* Brongn.

Tribe Zamieae

Subtribe: Microcycadinae

Microcycas. One species in Cuba; type: *M. calocoma* (Miquel) A. DC.

Subtribe: Zamiinae

Chigua. Two species in Colombia; type: *C. restrepoi* E. Stevenson

Zamia. About 65 species in the New World from Georgia, USA south to Bolivia; type: *Z. pumila* L.; see also *Z. furfuracea*

History

Modern knowledge about Cycads began in the 9th century with the recording by two Arab naturalists that the genus *Cycas* was used as a source of flour in India. Later, in the 16th century, Antonio Pigafetta, Fernao Lopez de Castanheda and Francis Drake found Cycas plants in the Moluccas, where the seeds were eaten. The first report of cycads in the New World was by Giovanni Lerio in his 1576 trip to Brazil, where he observed a plant named *ayrius* by the indigenous people; this species is now classified in the genus *Zamia*.

Cycads belonging to the genus *Encephalartos* were first described by Johann Georg Christian Lehmann in 1834. The name is derived from the Greek articles "en", meaning "in", "cephale", meaning "head", and "artos", meaning "bread".

Throughout the 18th-19th centuries, discoveries of several species were reported by numerous naturalist researchers and discoverers traveling throughout the world. One of the most notable researchers of cycads was American botanist C.J. Chamberlain whose work is noteworthy for the quantity of data and the novelty of his approach to studying cycads. His 15 years of travel throughout Africa, the Americas and Australia to observe cycads in their natural habitat resulted in his 1919 publication of *The Living Cycads* which remains current in its synthesis of taxonomy, morphology and reproductive biology of cycads, most of which was obtained from his original research. His 1940s monograph on the Cycadales, though never published (most likely because of his death) was never used by botanists. There are no other complete works on the cycads.

Uses

The generic name refers to the starch obtained from the stems which was used as food by some indigenous tribes. Tribal people grind and soak the nuts to remove the nerve toxin, making the food source generally safe to eat, although often not all the toxin is removed. In addition, consumers of bush meat may face a health threat as the meat comes from game which may have eaten cycad nuts and carry traces of the toxin in body fat.

There is some indication that the regular consumption of starch derived from cycads is a factor in the development of Lytico-Bodig disease, a neurological disease with symptoms similar to those of Parkinson's disease and ALS. Lytico-Bodic and its potential connection to cycasin ingestion is one of the subjects explored in Oliver Sacks' 1997 book *Island of the Colourblind.*

Distribution

Approximate world distribution of living Cycadales

Overall species diversity peaks at 17° 15"N and 28° 12"S, with a minor peak at the equator. There is therefore not a latitudinal diversity gradient towards the equator but towards the tropics. However, the peak in the northern tropics is largely due to *Cycas* in Asia and *Zamia* in the New World, whereas the peak in the southern tropics is due to *Cycas* again, and also to the diverse genus *Encephalartos* in southern and central Africa and *Macrozamia* in Australia. Thus the distribution pattern of cycad species with latitude appears to be an artifact of the geographical isolation of cycad genera, and is dependent on the remaining species in each genus that did not follow the extinction pattern of their ancestors. *Cycas* is the only genus that has a broad geographical range and can thus be used to infer that cycads tend to live in the upper and lower tropics. This is probably because these areas have a drier climate with relatively cool winters; while cycads require some rainfall, they appear to be partly xerophytic. Potted specimens are found and thrive in global locations such as Canada, Russia, Finland, Chile.

SPECIATION

There are no documented cases of sympatric speciation in cycads and allopatry appears to be the most common form of speciation in the group. This is difficult to study as they are long-lived plants, and so natural experiments have been investigated. One example is *Cycas seemannii,* which occurs only in Fiji, New Caledonia, Tonga and Vanuatu. Genetic diversity within populations was found to be significantly lower than between islands, suggesting that genetic drift is a likely mechanism for speciation, and is probably currently occurring between the isolated populations. Allopatry has also been proposed as the mechanism of speciation in *Dioon,* which predominantly occurs in Mexico. The many rivers that have shaped the region, and repeated glaciation and consequent disjunction, are thought to have been important in reproductive isolation not only in *Dioon* but in many other plant and animal taxa. Parapatric speciation may also have occurred, especially as cycads are pollinated by insects rather than by wind. As the range of the species grows, the individuals furthest apart are prevented from interbreeding as insects have relatively small ranges and will not pollinate between these plants. If sympatric speciation has occurred in cycads this would most likely be because of a host shift in pollinators, due to the very fact that cycads are uniformly dioecious.

Extinction

The probable former range of cycads can be inferred from their global distribution. For example, the family Stangeriaceae only contains three extant species, in Africa. Diverse fossils of this family have been dated to 135 mya, indicating that diversity may have been much greater before the Jurassic and late Triassic mass extinction events. However, the cycad fossil record is generally poor and little can be deduced about the effects of each mass extinction event on their diversity.

Instead, correlations can be made between the number of extant gymnosperms and angiosperms. It is likely that cycad diversity was affected more by the great angiosperm radiation in the mid-Cretaceous than by extinctions. Very slow cambial growth was first used to define cycads, and because of this characteristic the group could not compete with the rapidly growing, relatively short-lived angiosperms, which now number over 250,000 species, compared to the 947 remaining gymnosperms. It is surprising that the cycads are still extant, having been faced with extreme competition and five major extinctions. The ability of cycads to survive in relatively dry environments where plant diversity is generally lower, and their great longevity may explain their long persistence.

CONSERVATION

Encephalartos woodii is extinct in the wild, and all living specimens are clones of the type.

In recent years, many cycads have been dwindling in numbers and may face risk of extinction because of theft and unscrupulous collection from their natural habitats, as well as from habitat destruction.

23% of the 305 extant cycad species are either critically endangered or endangered, and 15% are vulnerable. Thus 38% of cycads are on the IUCN Red List (2004), and the other 62% are in the Least concern or Near Threatened category (i.e. not actually on the Red List), or are data deficient. This value has changed dramatically within the past few years; 46% of cycads were on the 1978 Red List, and this rose to 82% in 1997. This was largely due to the recent discovery of over 150 new species, disagreements about classification, and uncertainty. This has not been helpful for conservation planning for the group.

Zamia in the New World, *Cycas* in Asia and *Encephalartos* in Africa are the most threatened genera. This pattern reflects the pressures on species in these regions. At least two species, *Encephalartos woodii* and *Encephalartos relictus* (both from Africa), are confirmed extinct in the wild. Cycads are long-lived with infrequent reproduction, and most populations are small, putting them at risk of extinction from habitat destruction and stochastic environmental events. Regionally, Australian cycads are the least at risk, as they are locally common and habitat fragmentation is low. However, land management with fire is thought to be a threat to Australian species. African cycads are rare and are thought to be naturally decreasing due to small population sizes, and there is controversy over whether to let natural extinction processes act on these cycads.

All cycads are in the CITES appendix appearing under the heading Plant Kingdom and under three family names, Cycadaceae, Stangeriaceae and Zamiaceae. Some cycas are given below:

- *Cycas beddomei*
- *Stangeria eriopus*
- All *Ceratozamia*
- All *Chigua*
- All *Encephalartos*
- *Microcycas calocoma*

HORTICULTURE

A Sago Cycad (*Cycas revoluta*) growing in England as a houseplant.

Cycads can be cut up into pieces to make new plants, although the most environmentally responsible method is by direct planting of the seeds. Propagation by seeds is the preferred method of growth, and two unique risks to their germination exist. One is that the seeds have no dormancy, so that the embryo is biologically required to maintain growth and development, which means if the seed dries out, it dies. The second is that the emerging radicle and embryo can be very susceptible to fungal diseases in its early stages when in unhygienic or excessively wet conditions. Thus, many cycad growers pre-germinate the seeds in moist, sterile media such as vermiculite or perlite. However pre-germination is not necessary, and many report success by directly planting the seeds in regular potting soil. As with many plants, a combination of well-drained soil, sunlight, water and nutrients will help it to prosper. Although, because of their hardy nature, cycads do not necessarily require the most tender or careful treatment, they can grow in almost any medium, including soil-less ones. One of the most common cause of cycad death is from rotting stems and roots due to over-watering.

Some insects, particularly scale insects, some weevils and chewing insects can damage cycads, though the pests are susceptible to insecticides such as the horticulture soluble oil white oil. Sometimes bacterial preparations may be used to control insect infestation on cycads. However, when some of the mature plants prepare for reproduction, the presence of weevils have been shown to help accomplish pollination.

While the cycads have a reputation of slow growth, it is not always well-founded and some actually grow quite fast, achieving reproductive maturity in 2–3 years (as with some *Zamia* species), while others in 15 years (as with some *Cycas*, Australian *Macrozamia* and *Lepidozamia*).

PLANT SEXUALITY

Plant sexuality covers the wide variety of sexual reproduction systems found across the plant kingdom. This article describes morphological aspects of sexual reproduction of plants.

Among all living organisms, Flowers which are the reproductive structures of angiosperms, are the most varied physically and show the greatest diversity in methods of reproduction of all biological systems. A system of classification of flowering plants based on plant structures, since plants employ many different morphological adaptations involving sexual reproduction, flowers played an important role in that classification system. Later on Christian Konrad Sprengel (1793) studied plant sexuality and called it the "revealed secret of nature" and for the first time it was understood that the pollination process involved both biotic and abiotic interactions (Charles Darwin's theories of natural selection utilized this work to promote his idea of evolution). Plants that are not flowering plants (green alga, mosses, liverworts, hornworts, ferns, and gymnosperms) also have complex interplays between morphological adaptation and environmental factors in their sexual reproduction. The breeding system, or how the sperm from one plant fertilizes the ovals of another, is the single most important determinant of the mating structure of nonclonal plant populations. The mating structure or morphology of the flower parts and their arrangement on the plant in turn controls the amount and distribution of genetic variation, a central element in the evolutionary process

History

Unlike animals, plants are immobile and cannot seek out sexual partners for reproduction. The first plants used abiotic means to transport sperm for reproduction, utilizing water and wind. The first plants were aquatic and released sperm freely into the water to be carried by the currents. As plants moved onto land they used a thin film of water or water droplets like liverworts and ferns, in which mobile sperm swam from the male reproduction organs to the female organs. As plants became more complex and developed vascular systems enabling them to grow taller, they used alternation of generations like in ferns or the wind to move spores. In the Paleozoic era progymnosperms reproduced by using spores dispersed on the wind, 350 million years ago the seed plants evolved, including seed ferns, conifers and cordaites all were gymnosperms. Pollen grains, the male gametophyte, developed for protection of the sperm during the process of transfer from male to female parts. It is believed that insects feed on the pollen and plants evolved to use insects to actively carry pollen from one plant to the next. Seed producing plants, which include the angiosperms and the gymnosperms, have hetromorphic alternation of generations with large sporophytes containing much reduced gametophytes. Angiosperms have distinctive reproductive organs called flowers with carpels and the gametophyte is greatly reduced to a female embryo sac with as few as eight cells and the male gametophyte develop from the pollen grains. The sperm of seed plants are non motile except for two older groups of plants the Cycadophyta and the Ginkgophyta which have flagellated sperm.

The flowers of angiosperms are determinate shoots that have sporophylls. The parts of flowers are named by scientists and show great variation in shape, these flower parts include sepals, petals, stamens and carpels. As a group the sepals form the calyx and as a group the petals form the corolla, together the corolla and the calyx is called the perianth. The stamens collectively are called the androecuim and the carpels collectively are called the gynoecium.

The complexity of the systems and devices used by plants to achieve sexual reproduction has resulted in botanists and evolutionary biologists using numerous terms to describe physical structures and functional strategies. A variety of terms used to describe the modes of sexuality at different levels in flowering plants. This list is reproduced here , generalized to fit more than just plants that have flowers, and expanded to include other terms and more complete definitions.

The Alder is monoecious. Shown here: maturing male flower catkins on right, last year's female catkins on left.

Individual reproductive unit (a flower in angiosperms)

- *Bisexual:* or perfect flowers have both male (androecium) and female (gynoecium) reproductive structures, including stamens, carpels, and an ovary. Flowers that contain both androecium and gynoecium are called androgynous or hermaphroditic. Examples of plants with perfect or bisexual flowers include the lily, rose, and most plants with large showy flowers, though a perfect flower does not have to have petals or sepals. Other terms widely used are hermaphrodite, monoclinous, and synoecious. A complete flower is a perfect flower with petals and sepals.

- *Unisexual:* Reproductive structure that is either functionally male or functionally female. In angiosperms this condition is also called diclinous, imperfect or incomplete.

Adaptations

Plants with wind pollinated flowers tend to have flowers without petals or sepals. Typically large amounts of pollen are produced and pollination often occurs early in the growing season before leaves can interfere with the dispersal of the pollen. Many trees and all grasses and sedges are wind pollinated, as such they have no need for large fancy flowers. In plants that use insects or other animals to move pollen from one flower to the next, plants have developed greatly modified flower parts to attract pollinators and to facilitate the movement of pollen from one flower to the insect and from the insect back to the next flower. Plants have a number of different means to attract pollinators including color, scent, heat, nectar glands, eatable pollen and flower shape. Along with modifications involving the above structures two other conditions play a very important role in the sexual reproduction of flowering plants, the first is timing of flowering and the other is the size or number of flowers produced. Often plant species have a few large, very showy flower while others produce many small flower, often flowers are collected together into large inflorescences to maximize their visual effect, becoming more noticeable to passing by pollinators. Flowers are attraction strategies and sexual expressions are functional strategies used to produce the next generation of plants, with pollinators and plants having co-evolved, often to some extraordinary degrees, very often mutually benefiting both.

The largest family of flowering plants is the orchids (Orchidaceae), estimated by some specialists to include up to 35,000 species, which often have highly specialized flowers used to attract insects and facilitate pollination. The stamens are modified to produce pollen in clusters called pollinium, which are attached to insects when crawling into the flower. The flower shapes are modified to force insects to pass by the pollen, which is "glued" to the insect. Some orchids are even more highly specialized, with flower shapes that mimic the shape of insects to attract them to 'mate' with the flowers, a few even have scents that mimic insect pheromones.

Another large group of flowering plants is the Asteraceae or sunflower family with close to 22,000 species which also have highly modified inflorescences that are flowers collected together in heads composed of a composite of individual flowers called florets. Heads with florets of one sex, when the flowers are pistillate or functionally staminate, or made up of all bisexual florets, are called homogamous and can include discoid and liguliflorous type heads. Some radiate heads may be homogamous too. Plants with heads that have florets of two or more sexual forms are called heterogamous and include radiate and disciform head forms, though some radiate heads may be heterogamous too.

Individual Plant Sexuality

Many plants have complete flowers that have both male and female parts, others only have male or female parts and still other plants have flowers on the same plant that are a mix of male and female flowers. Some plants even have mixes that include all three types of flowers, where some flowers are only male, some are only female and some are both male and female. A distinction needs to be made between arrangements of sexual parts and the expression of sexuality in single plants verses the species. Some plants also undergo what is called Sex-switching, like *Arisaema triphyllum* which express sexual differences at different stages of growth. In some arums smaller plants produce all or mostly male flowers and as plants grow larger over the years the male flowers are replaced by more female flowers on the same plant. *Arisaema triphyllum* thus covers a multitude of sexual conditions in its life time; from nonsexual juvenile plants to young plants that are all male, as plants grow larger they have a mix of both male and female flowers, to large plants that have mostly female flowers Other species have plants that produce more male flowers early in the year and as plants bloom later in the growing season they produce more female flowers. In plants like *Thalictrum dioicum* all the plants in the species are ether male or female.

Specific terms are used to describe the sexual expression of individual plants within a population.

- *Androecious* - plants producing male flowers only, produce pollen but no seeds, the male plants of a Dioecious species.
- *Dioecious* - having unisexual reproductive units with male and female plants. (flowers, conifer cones, or functionally equivalent structures) occurring on different individuals; from Greek for "two households". Individual plants are not called dioecious: they are either gynoecious (female plants) or androecious (male plants).
- *Gynoecious* - plants producing female flowers only, produces seeds but no pollen, the female of a Dioecious species. In some plant species or populations all individuals are gynoecious with non sexual reproduction used to produce the next generation.
- *Hermaphrodite* - A plant that has only bisexual reproductive units (flowers, conifer cones, or functionally equivalent structures). In angiosperm terminology a synonym is monoclinous from the Greek "one bed".
- *Monoecious* - having separate male and female reproductive units (flowers, conifer cones, or functionally equivalent structures) on the same plant; from Greek for "one household". Individuals bearing separate flowers of both sexes at the same time are called simultaneously or synchronously monoecious. Individuals that bear flowers of one sex at one time are called consecutively monoecious; Plants may first have single sexed flowers and then later have flowers of the other sex. Protoandrous describes individuals that function first as males and then change to females; protogynous describes individuals that function first as females and then change to males.
- *Subdioecious* - a tendency in some dioecious species to produce monoecious plants. The population produces normally male or female plants but some are hermaphroditic, with female plants producing some male or hermaphroditic flowers or vise versa. The condition is thought to represent a transition between hermaphroditism and dioecy.

- *Gynomonoecious* - has both hermaphrodite and female structures.
- *Andromonoecious* - has both hermaphrodite and male structures.
- *Subandroecious* - plant has mostly male flowers, with a few female or hermaphrodite flowers.
- *Subgynoecious* - plant has mostly female flowers, with a few male or hermaphrodite flowers.
- *Trimonoecious* (*polygamous*) - male, female, and hermaphrodite structures all appear on the same plant.
- *Diclinous* ("two beds"), an angiosperm term, includes all species with unisexual flowers, although particularly those with *only* unisexual flowers, i.e. the monoecious and dioecious species.

Holly (*Ilex aquifolium*) is dioecious: (above) shoot with flowers from male plant; (top right) male flower enlarged, showing stamens with pollen and reduced, sterile stigma; (below) shoot with flowers from female plant; (lower right) female flower enlarged, showing stigma and reduced, sterile stamens with no pollen.

PLANT POPULATION

Most often plants show uniform strategies across the species or in populations in their sexual expression and specific terms are used to describe the sexual expression of the species or population.

- *Hermaphrodite* - only hermaphrodite plants with flowers that have both male and female parts.
- *Monoecious* - only monoecious plants, that is plants have separate male and female flowers on the same plant. A plant species where the male and female organs are found in different flowers on the same plant. These plants are often wind pollinated. Examples of monoecious plants include corn, birch and pine trees, and most fig species.

- *Dioecious* - only dioecious plants, all plants are either female or male.
- *Gynodioecious* - both female and hermaphrodite plants present. In some plants, strictly female plants are produced by the degeneration of the tapetum, a shell-like structure in the anther of a flower where the pollen cells form, producing male sterility often regulated by environmental factors like temperature, photo period or water availability.
- *Androdioecious* - both male and hermaphrodite plants present.
- *Polygamous* - when there is a mix of hermaphrodite and unisexual plants in the natural population.
- *Subdioecious* - population of unisexual (dioecious) plants, with monoecious individuals too.
- *Trioecious* - sometimes used in place of subdioecious when male, female, and hermaphrodite plants are more equally mixed with in the same population.

About 11% of all angiosperms are strictly dioecious or monoecious, Intermediate forms of sexual dimorphism, including gynodioecy and androdioecy, represent 7% of the species examined of a survey of 120,000 plant species. In the same survey, 10% of the species contain both unisexual and bisexual flowers.

The majority of plant species use allogamy - also called cross-pollination, as a means of breeding. Many plants are self fertile and the male parts can pollinate the female parts of the same flower and/or same plant. Some plants use a method known as self-incompatibility to promote outcrossing. In these plants, the male organs cannot fertilize the female parts of the same plant, other plants produce male and female flowers at different times to promote outcrossing. Some plants have bisexual flowers but the pollen is produced before the stigma of the same flower is receptive of pollen, this promotes out crossing by greatly limiting self pollination and these types of flowers are called protandrous.

Flower Morphology

A species such as the ash tree (*Fraxinus excelsior* L.), demonstrates the possible range of variation in morphology and functionality exhibited by flowers with respect to gender. Flowers of the ash are wind-pollinated and lack petals and sepals. Structurally, the flowers may be either male or female, or even hermaphroditic, consisting of two anthers and an ovary. A male flower can be morphologically male or hermaphroditic, with anthers and a rudimentary gynoecium. Ash flowers can also be morphologically female, or hermaphroditic and functionally female.

Angiosperms

It is thought that flowering plants evolved from a common hermaphrodite ancestor, and that dioecy evolved from hermaphroditism. Hermaphroditism is very common in flowering plants; over 85% are hermaphroditic, whereas only about 6-7% are dioecious and 5-6% are monoecious.

A fair degree of correlation (though far from complete) exists between dioecy/sub-dioecy and plants that have seeds dispersed by birds (both nuts and berries). It is hypothesized that the concentration of fruit in half of the plants increases dispersal efficiency; female plants can produce a higher density of fruit as they do not expend resources on pollen production, and the dispersal agents (birds) need not waste time looking for fruit on male plants. Other correlations with dioecy include: tropical distribution, woody growth form, perenniality, fleshy fruits, and small, green flowers. can be used to alter flower and plant sexuality, in cucumbers ethephon is used to delay staminate flowering and transforms monoecious lines into all-pistillate or female lines. Gibberellins also increase maleness in cucumbers. Cytokinins have been used in grapes that have undeveloped pistils to produce functional female organs and seed formation.

CYCAS

Cycas is the type genus and the only genus currently recognised in the cycad family Cycadaceae. About 95 species are currently accepted. The best-known species is *Cycas revoluta*, widely cultivated under the name "Sago Palm" or "King Sago Palm" due to its palm-like appearance although it is not a true palm. The generic name comes from Greek *Koikas,* and means "a kind of palm".

The genus is native to the Old World, with the species concentrated around the equatorial regions. It is native to eastern and southeastern Asia including the Philippines with 10 species (9 of which are endemic), eastern Africa (including Madagascar), northern Australia, Polynesia, and Micronesia. Australia has 26 species, while the Indo-Chinese area has about 30. The northernmost species (*C. revoluta*) is found at 31°N in southern Japan. The southernmost (*C. megacarpa*) is found at 26°S in southeast Queensland, Australia.

The plants are dioecious, and the family Cycadaceae is unique among the cycads in not forming seed cones on female plants, but rather a group of leaf-like structures each with seeds on the lower margins, and pollen cones on male individuals.

The caudex is cylindrical, surrounded by the persistent petiole base. Most species form distinct branched or unbranched trunks but in some species the main trunk can be subterranean with the leaf crown appearing to arise directly from the ground. The leaves are pinnate (or more rarely bipinnate) and arranged spirally, with thick and hard keratinose. The leaflets are articulated, have midrib but lack secondary veins. Megasporophylls are not gathered in cones.

Often considered a living fossil, the earliest fossils of the genus *Cycas* appear in the Cenozoic although *Cycas*-like fossils that may belong to Cycadaceae extend well into the Mesozoic. *Cycas* is not closely related to other genera of cycads, and phylogenetic studies have shown that Cycadaceae is the sister-group to all other extant cycads.

The plant takes several years to grow, sexual reproduction takes place after 10 years of exclusive vegetative growth.

Species

Cycas aculeata

Cycas angulata

Cycas annaikalensis

Cycas apoa

Cycas arenicola

Cycas armstrongii

Cycas arnhemica

Cycas badensis

Cycas balansae

Cycas basaltica

Cycas beddomei

Cycas bifida

Cycas bougainvilleana

Cycas brachycantha

Cycas brunnea

Cycas cairnsiana

Cycas calcicola

Cycas campestris

Cycas candida

Cycas canalis

Cycas chamaoensis

Cycas changjiangensis

Cycas chevalieri

Cycas circinalis

Cycas clivicola

Cycas collina

Cycas condaoensis

Cycas conferta

Cycas couttsiana

Cycas curranii

Cycas debaoensis

Cycas desolata

Cycas diannanensis

Cycas dolichophylla

Cycas edentata

Cycas elephantipes

Cycas elongata

Cycas falcata

Cycas fairylakea

Cycas ferruginea

Cycas fugax

Cycas furfuracea

Cycas guizhouensis

Cycas hainanensis

Cycas hoabinhensis

Cycas hongheensis

Cycas inermis

Cycas javana

Cycas lanepoolei

Cycas lindstromii

Cycas litoralis

Cycas maconochiei

Cycas macrocarpa
Cycas media
Cycas megacarpa
Cycas micholitzii
Cycas micronesica
Cycas multipinnata
Cycas nathorstii
Cycas nongnoochiae
Cycas ophiolitica
Cycas orientis
Cycas pachypoda
Cycas panzhihuaensis
Cycas papuana
Cycas pectinata
Cycas petraea
Cycas platyphylla
Cycas pranburiensis
Cycas pruinosa
Cycas revoluta
Cycas riuminiana
Cycas rumphii
Cycas schumanniana
Cycas scratchleyana
Cycas seemafaux
Cycas segmentifida
Cycas semota
Cycas sexseminifera
Cycas siamensis
Cycas silvestris
Cycas simplicipinna
Cycas spherica
Cycas szechuanensis
Cycas taitungensis
Cycas taiwaniana
Cycas tanqingii
Cycas tansachana
Cycas thouarsii
Cycas tropophylla
Cycas tuckeri
Cycas wadei
Cycas xipholepis
Cycas yorkiana
Cycas yunnanensis
Cycas zambalensis
Cycas zeylanica

9

Pteridomania

INTRODUCTION

Pteridomania and Fern-Fever are terms for the Victorian era craze of fern collecting and fern motifs in decorative art including pottery, glass, metals, textiles, wood, printed paper, and sculpture "appearing on everything from christening presents to gravestones and memorials."

A plate from *The Ferns of Great Britain and Ireland*, a book from the era of pteridomania.

The term pteridomania, which means "Fern Madness" or "Fern Craze", derives from the group of plant to which ferns belong, the *Pteridophytes*, and the word "mania" meaning "madness" or "craze". It was coined in 1855 by Charles Kingsley in his book *Glaucus, or the Wonders of the Shore* in which he wrote:

> Your daughters, perhaps, have been seized with the prevailing 'Pteridomania'...and wrangling over unpronounceable names of species (which seem different in each new Fern-book that they buy)...and yet you cannot deny that they find enjoyment in it, and are more active, more cheerful, more self-forgetful over it, than they would have been over novels and gossip, crochet and Berlin-wool.

According to one researchers:

Although the main period of popularity of ferns as a decorative motif extended from the 1850s until the 1890s, the interest in ferns had really begun in the late 1830s when the British countryside attracted increasing numbers of amateur and professional botanists (male and female). New discoveries were published in periodicals, particularly 'The Phytologist' which first appeared in 1841. Ferns proved to be a particularly fruitful group of plants for new records because they had been relatively little studied compared with flowering plants. Also, they were most diverse and abundant in the wilder, wetter, western and northern parts of Britain which were becoming more accessible through the development of better roads and, subsequently, in the late 1840s and 1850s through the development of a railway network

COLLECTION AND CULTIVATION

The Wardian case, a forerunner of the terrarium, helped protect Victorian fern collections from the air pollution of the era.

The collection of ferns drew enthusiasts from different social classes and it is said that "even the farm labourer or miner could have a collection of British ferns which he had collected in the wild and a common interest sometimes brought people of very different social backgrounds together."

For some a fashionable hobby and for others a more serious scientific pursuit, fern collecting became commercialized with the sale of merchandise for fern collectors. Equipped with *The Ferns of Great Britain and Ireland* or one of the many other books sold for fern identification, collectors sought out ferns from dealers and in their native habitats across the British Isles and beyond. Fronds were pressed in albums for display in homes. Live plants were also collected for cultivation in gardens and indoors. Nurseries provided not only native species but exotic species from the Americas and other parts of the world.

The Wardian case, a forerunner of the modern terrarium, was invented about 1829 by a physician to protect his ferns from the air pollution of 19th century London. Wardian cases soon became features of stylish drawing rooms in Western Europe and the United States and helped spread the fern craze and the craze for growing orchids that followed. Ferns were also cultivated in fern houses (greenhouses devoted to ferns) and in outdoor ferneries.

Besides approximately seventy native British species and natural hybrids of ferns, horticulturalists of this era were very interested in so-called "monstrosities", odd variants of wild species. From these they selected hundreds of varieties for cultivation. *Polystichum setiferum, Athyrium filix-femina* and *Asplenium scolopendrium,* for example, each yielded about three hundred different varieties

DECORATIVE ART

Circa 1860 English-made Victorian majolica pitcher with fern designs.

Fern motifs first became conspicuous at 1862 International Exhibition and remained popular "as fond symbol of pleasurable pursuits" until the turn of the century.

As fern fronds are somewhat flat they were used for decoration in ways that many other plants could not. They were glued into collectors' albums, affixed to three dimensional objects, used as stencils for "spatter-work", inkled and pressed into surfaces for nature printing, and so forth.

Fern pottery patterns were introduced by Wedgwood, Mintons Ltd, Royal Worcester, Ridgeway, George Jones and others with various shapes and styles of decoration including majolica. A memorial to Sir William Jackson Hooker, Director of the Royal Botanic Gardens, Kew was made commissioned from Josiah Wedgwood and Sons and erected in Kew Church in 1867 with jasperware panels with applied sprigs representing exotic ferns. A copy was presented to what is now the Victoria and Albert Museum where it may still be seen.

While realistic depictions of ferns were especiall favored in the decorative arts of this period, "Even when the representation was stylised such as was common on engraved glass and metal, the effect was still recognisably 'ferny'."

Other Species

Selaginella and *Lycopodiopsida* and other fern ally plants were also collected and represented on decorative objects.

EFFECTS ON NATIVE POPULATIONS

The zeal of Victorian collectors led to significant reductions in the wild populations of a number of the rarer species. Oblong Woodsia came under severe threat in Scotland, especially in the Moffat Hills. This area once had the most extensive UK populations of the species but there now remain only a few small colonies whose future remains under threat. The related Alpine Woodsia suffered a similar fate, although the risks were not all to the plants. John Sadler, later a curator of the Royal Botanic Garden Edinburgh, nearly lost his life obtaining a fern tuft on a cliff near Moffat and a botanical guide called William Williams died collecting Alpine Woodsia in Wales in 1861. His body was found at the foot of the cliff where Edward Lhwyd had first collected the species nearly two centuries earlier.

The Killarney Fern, considered to be one of Europe's most threatened plants and once found on Arran, was thought to be extinct in Scotland due to the activities of 19th century collectors, but the species has since been discovered on Skye in its gametophyte form. Dickie's Bladder-fern, which was discovered growing on base-rich rocks in a sea cave (known locally as a "yawn") on the coast of Kincardineshire in 1838. By 1860 the original colony had been apparently been extirpated, although the species has recovered and today there is a population of more than 100 plants there, where it grows in a roof fissure.

While interest in ferns may have increased to some degree outside United Kingdom of Great Britain and Ireland, it did not do so to the same extent anywhere else. According to someone at the Rockland Botanical Garden in Mertztown, Pennsylvania:

The craze seemed to have passed America by – most likely because these same species in America are essentially free of these "freaky" abnormal forms. It may also be do to the fact the American botanists have been for the most part more interested in unraveling the complexities of the species involved in the fern complexes such as Asplenium, Dryopteris, and Botrychium.

Nevertheless, the American Fern Society was established in 1893 and now has over 900 members worldwide. The society is based at the Indiana University and counts itself as "one of the largest international fern clubs in the world." William Ralph Maxon served repeatedly as the society president.

The Dorrance H. Hamilton Fernery at the Morris Arboretum of the University of Pennsylvania is only remaining freestanding Victorian fernery in North America. It has a curved Victorian-style glass roof and turned 100 years old in 1999. Designed by John Morris, the arboretum's namesake, the Fernery is said to embody "some of the many passions of the Victorians: a love of collecting, a veneration of nature, and the fashion of romantic gardens ... its filigree roof sparkling in sunlight.

10

Brackens

INTRODUCTION

Brackens (*Pteridium*) are a genus of about ten species of large, coarse ferns, in the family Dennstaedtiaceae. The genus has probably the widest distribution of any fern genus in the world, being found on all continents except Antarctica and in all environments except for hot and cold deserts. Therefore it is considered to have a cosmopolitan distribution. In the past, the genus was commonly treated as having only one species, *Pteridium aquilinum*, but the recent trend is to subdivide it into several species.

It can also have a major impact on archaeological remains, by disturbing or destroying below-ground archaeological interest. It also obscures archaeological sites which could lead to their inadvertent damage.

The word bracken is of Old Norse origin, related to the Swedish word *bräken*, meaning fern.

Evolutionarily, bracken may be considered to be one of the most successful ferns. It is also one of the oldest, with fossil records of over 55 million years old having been found. The plant sends up large, triangular fronds from a wide-creeping underground rootstock, and may form dense thickets. This rootstock may travel a metre or more underground between fronds. The fronds may grow up to 2.5 m (8 ft) long or longer with support, but typically are in the range of 0.6-2 m (2-6 feet) high. In cold environments bracken is winter-deciduous, and, as it requires well-drained soil, is generally found growing on the sides of hills.

It is a herbaceous perennial plant, deciduous in winter. The fronds are produced singly from an underground rhizome, and grow to 1-3 m tall; the main stem is up to 1 cm diameter at the base. The rhizomes typically grow to a depth of 50cm, although in some soils this may extend to more than a metre.

The spores used in reproduction are produced on the underneath of the leaf in structures found on the edges of the leaf called sorus. The linear pattern of these is different to other ferns which are circular and towards the centre.

Pteridium aquilinum (Bracken or Common Bracken) is the most common species with a cosmopolitan distribution, occurring in temperate and subtropical regions throughout much of the world, including most of Europe, Asia, and North America in the Northern Hemisphere, and Australia, New Zealand and northern South America in the Southern Hemisphere. It is a prolific and abundant plant in the highlands of Great Britain. It is limited to altitudes of below 600 metres in the UK, does not like extreme cold temperatures, poorly drained Marshes or Fen. It causes such a problem of invading pastureland that at one time the British government had an eradication program. Special filters have even been used on some British water supplies to filter out the bracken spores. . NBN distribution map for the United Kingdom.

It has been observered growing in soils from pH 2.8 to 8.6. Exposure to cold or high pH inhibits its growth.

FUNGI ASSOCIATIONS

Woodland fungi can be found growing under the bracken canopy, for example *Mycena epipterygia*. *Camarographium stephensii* is host specific to the dead stems.

Other Plant Associations

Allelopathy: Bracken fern is known to produce and release allelopathic chemicals, which is an important factor in its ability to dominate other vegetation, particularly in regrowth after fire. Herb and tree seedling growth may be inhibited even after bracken fern is removed, apparently because active plant toxins remain in the soil.

Brackens substitute the characteristics of a woodland canopy, and are important for giving shade to european plants such as common bluebell and wood anemone, where the woodland does not exist. These plants are intolerant to stock trampling. Dead bracken provides a warm microclimate for development of the immature stages. Climbing corydalis,wild gladiolus and chickweed wintergreen also seem to benefit from the conditions found under bracken stands.

The high humidity helps mosses survive underneath including *Campylopus flexuosus, Hypnum cupressiforme, Polytrichum commune, Pseudoscelopodium purum* and *Rhytidiadelphus squarrosus.*

ANIMAL SPECIES THAT USE BRACKEN

Brackens of the northern hemisphere are used as food plants by the larvae of some Lepidoptera species including Dark Green Fritillary, Dot Moth, High Brown Fritillary, Gold Swift, Map-winged Swift, Pearl-bordered Fritillary, Orange Swift, Small Angle Shades, Small Pearl-bordered Fritillary. They also form an important ecological partnership with plants such as violet and cow-wheat (*Melampyrum pratense*) for various Boloria Fritillary species.

It is also a favoured haunt of the sheep tick *Ixodes ricinus* which can carry Lyme Disease.

Between 27 to 40 invertebrates (including nine moths) in the UK feed on bracken. These include the sawfly, a plant hopper (*Dytroptis pteridis*), the map-winged swift moth caterpillar, brown silver-line moth caterpillar (*Petrophora chlorosata*) and *Paltodora cytisella*. The numbers feeding on the bracken increase as the season progresses due to the decreasing levels of toxin, and the production of nectaries in the spring, food for ants which in turn may kill any herbivorous insects in the vicinity.

Some birds such as the whinchat and the nightjar use bracken as their preferred habitats. The nightjar may lay its eggs on the bare ground under the bracken. The skylark often nests in bracken and uses it for cover. Other birds known to nest in or beneath bracken include the willow warbler (it will also use bracken to construct its nest), the tree pipit, the yellowhammer, the ring ouzel, the woodcock and the twite.

The European adder can be found basking on bracken, the colour of their skin concealing them.

Bracken fiddleheads (the immature, tightly curled emerging fronds) have been considered edible by many cultures throughout history, and are still commonly used today as a foodstuff. Bracken fiddleheads are either consumed fresh (and cooked) or preserved by salting, pickling, or sun drying. In Korea, where they are called *gosari namul* , they are a typical ingredient in the mixed rice dish called *bibimbap*.

Both fronds and rhizomes have been used to brew beer, and the rhizome starch has been used as a substitute for arrowroot. Bread can be made out of dried and powered rhizomes alone or with other flour. American Indians cooked the rhizomes, then peeled and ate them or pounded the starchy fiber into flour. In Japan, starch from the rhizomes is used to make confections.

Bracken is called *wiwnunmí útpas* 'huckleberry's blanket' by the Umatilla Indians of the Columbia River in the United States Northwest. The fronds were used to cover a basket full of huckleberries in order to keep them fresh.

The Maori of New Zealand used the rhizomes of *P. esculentum* (*aruhe*) as a staple food, especially for exploring or hunting groups away from permanent settlements; much of the widespread distribution of this species in present-day New Zealand is in fact a consequence of prehistoric deforestation and subsequent tending of *aruhe* stands on rich soils (which produced the best rhizomes). The rhizomes were air-dried so that they could be stored and became lighter; for consumption, they were briefly heated and then softened with a *patu aruhe* (rhizome pounder); the starch could then be sucked from the fibers by each diner, or collected if it were to be prepared for a larger feast. *Patu aruhe* were significant items and several distinct styles were developed (McGlone *et al.* 2005).

Bracken has also been used as a form of herbal remedy. Powdered rhizome has been considered particularly effective against parasitic worms. American Indians ate raw rhizomes as a remedy for bronchitis.

In East Asia, *Pteridium aquilinum* (fernbrake or bracken fiddleheads) is eaten as a vegetable, called *warab*) in Japan, *gosari* in Korea, and *juécài* in China and Taiwan. In Korea, a typical *banchan* (small side dish) is *gosari-namul* that consists of prepared fernbrake that has been sauteed. It is a component of the popular dish *bibimbap*.

Bracken has been shown to be carcinogenic in some animals and is thought to be an important cause of the high incidence of stomach cancer in Japan. It is currently under investigation as a possible source of new insecticides.

Uncooked bracken contains the enzyme thiaminase, which breaks down thiamine. Eating excessive quantities of bracken can cause beriberi, especially in creatures with simple stomachs. Ruminants are less vulnerable because they synthesize thiamine. It was traditionally used for animal bedding, which later broke down to a rich mulch which could be used as fertiliser. Other uses were as packing material for products such as earthenware, as a fuel, as a form of thatch. The ash was used for degreasing woolen cloth. The ash of bracken fern was used in making *forest glass* in Central Europe from about 1000 to 1700.

Poisoning

The plant is carcinogenic to animals such as mice, rats and cattle when ingested, although they will usually avoid it unless nothing else is available. Young stems are used as a vegetable in Japan, leading some researchers to suggest a link between consumption and higher stomach cancer rates The spores have also been implicated as a carcinogen. Danish scientist Lars Holm

Rasmussen released a study in 2004 showing that the carcinogenic compound in bracken, ptaquiloside or PTQ, can leach from the plant into the water supply, which may explain an increase in the incidence of gastric and oesophageal cancers in bracken-rich areas

In cattle, bracken poisoning can occur in both an acute and chronic form, acute poisoning being the most common. In pigs and horses bracken poisoning induces vitamin B deficiency Poisoning usually occurs when there is a shortage of available grasses such as in drought or snowfalls.

It damages blood cells and destroys thiamine (Vitamin B1). This in turn causes beriberi, a disease linked to nutritional deficiency.

Control

Various techniques are recommended by Natural England to control bracken either individually or in combination:

- Cutting - once or twice a year, for at least three years
- Crushing - using heavy rollers, again for at least three years
- Livestock treading - during winter, encouraging livestock to bracken areas with food. They trample the developing plants and allow frost to penetrate the rhizomes. Livestock should be removed in the spring to prevent them being poisoned.
- Burning - useful for removing the litter, but may be counter-productive as bracken is considered to be a fire adapted species
- Ploughing - late in the season followed by sowing seed.
- Herbicide - Asulam is selective for ferns, and Glyphosate is not but has the advantage that the effects can be seen soon after application. They are applied when the fronds are fully unfurled to ensure that the chemical is fully absorbed. Natural England recommends that only Asulam can be sprayed aerially, Glyphosate requires spot-treatment e.g. using a weedwiper or knapsack spray.
- Allowing plants to grow in its place, e.g., the establishment of woodland, causes shade that inhibits bracken growth.

DENNSTAEDTIACEAE

Dennstaedtiaceae is one of fifteen families in the order Polypodiales, the most derived families within monilophytes (ferns). It includes the world's most abundant fern, *Pteridium aquilinum* (bracken). Members of the order generally have large, highly divided leaves and have either small, round intramarginal sori with cup-shaped indusia (e.g. *Dennstaedtia*) or linear marginal sori with a false indusium formed from the reflexed leaf margin (e.g. *Pteridium*). The morphological diversity among members of the order has confused past taxonomy, but recent molecular studies have supported the monophyly of the order and the family The reclassification of Dennstaedtiaceae and the rest of the monilophytes was published in 2006 , so most of the available literature is not updated.

Characteristics

- Terrestrial or scrambling over other vegetation—scandent
- Rhizomes long-creeping, occasionally short-creeping
- Rhizomes bear jointed hairs
- Rhizomes often siphonostelic or polystelic
- Petioles often with epipetiolar buds
- Petioles usually with gutter-shaped vascular strand (open end adaxial)
- Petiole pubescent of glabros
- Blades often large
- Blades often 2-3 pinnate, can be 1-4 pinnate or more divided
- Inducement of hairs, no scales
- Veins free, or forked, or pinnate, rarely vanishing
- Sori near the margin, or submarginal, sometimes fused with the blade to form a cup or pouch or obscured in a recurved portion of the blade margin
- Sori mostly linear, may be discrete
- Indusia linear or cup shaped at blade margin, or reflexed over sori

- Sporangium stalk with 1-3 rows of cells
- Spore tetrahedral and trilete, or reniform and monolete
- Gametophyte green, cordate

Characteristics described by Smith et al., and Judd et al.

Distribution of Genera

Generally, the family is pantropical, but due to the distribution of *Pteridium* (the most wide spread fern genus), Dennstaedtiaceae can be found worldwide . *Pteridium* is a well adapted early successional genus, generally described as a weed because of its ease of spread. The spore is light and robust, so it can travel relatively far and colonize open, disturbed environments easily. *Dennsteadtia* is mostly tropical to warm-temperate, but not well represented in the Amazon or Africa. *Oenotrichia* is in New Caledonia. *Leptolepia* is in New Zealand, Queensland (Australia), and in New Guinea. *Microlepia* is in the Asiatic-Pacific. *Paesia* occurs in tropical America, Asia, and the western Pacific. *Hypolepis* is tropical and south-temperate. *Blotiella* is strongly centered in Africa. *Histiopteris* is generally Malesian, with one pantropic to south-temperate species.

History of Classification

Dennstaedtiaceae was previously considered the only family the order Dennstaedtiales. Dennstaedtiaceae now contains the previously defined families Monachosoraceae Ching, Pteridiaceae Ching, and Hypolepidaceae Pic. Serm. . Before Smith's classification in 2006, Dennstaedtiaceae was a poly- and para- phyletic family containing genera that now are classified within Lindsaeaceae and Saccolomataceae, and with the family Monachosoraceae arising from within the Dennstaedtiaceae clade. The nonmonophyletic nature of Dennstaedtiaceae (pre-2006 classification) was proved and supported by multiple molecular studies. Dennstaedtiaceae as now classified is supported as monophyletic, but the relation of the genera within the family have not yet been fully clarified.

Interesting Species within Dennstaedtiaceae

Dennsteadtiaceae species and genera are usually known for their weedy nature (i.e. *Pteridium* spp., *Hypolepis* spp., *Paesia* spp.), but some species are grown ornamentally (*Blotiella* spp., *Dennstaedtia* spp., *Hypolepis* spp., *Microlepia* spp.)

The fiddleheads/crosiers of *Pteridium aquilinum* have been known to be eaten, but they contain carcinogens, so this practice is not prevalent.

The rhizomes of *Pteridium esculentum* were consumed by the Maori during their settlement of New Zealand in the 13th century, but no longer are a part of the Maori diet . The rhizomes of *Pteridium esculentum* contain about 50% starch when they grow in loose rich soil, at relatively deep depths[The rhizomes were a staple in the diet because once dried, the rhizomes were very light (perfect for traveling) and would keep for about a year as long as they remained dry . The leaves and spores of the *Pteridium esculentum* are associated with toxins and carcinogens, and have been known to cause stock (cattle, sheep, horses, pigs) to sicken.

EQUISETUM

Equisetum is a genus of vascular plants that reproduce by spores rather than seeds. The genus includes 15 species commonly known as horsetails and scouring rushes. It is the only living genus in class Equisetopsida, formerly of the division Equisetophyta (Arthrophyta in older works), though recent molecular analyses place the genus within the ferns (Pteridophyta). Other classes and orders of Equisetopsida are known from the fossil record, where they were important members of the world flora during the Carboniferous period.

Vegetative stem: N = node, I = internode, B = branch in whorl, L = leaves.

The name horsetail, often used for the entire group, arose because the branched species somewhat resemble a horse's tail, the name *Equisetum* being from the Latin *equus*, "horse", and *seta*, "bristle". Ironically *Equisetum* is poisonous to horses. Other names include candock (applied to branching species only), and scouring-rush (applied to the unbranched or sparsely branched species). The latter name refers to the plants' rush-like appearance; the stems were used for scouring cooking pots in the past (due to them being coated with abrasive silica).

Distribution

The genus is near-cosmopolitan, being absent only from Australasia and Antarctica. They are perennial plants, either herbaceous, dying back in winter (most temperate species) or evergreen (some tropical species, and the temperate species *Equisetum hyemale*, *E. scirpoides*, *E. variegatum* and *E. ramosissimum*). They mostly grow 0.2-1.5 m tall, though *E. telmateia* can exceptionally reach 2.5 m, and the tropical American species *E. giganteum* 5 m, and *E. myriochaetum* 8 m.

Anatomy

In these plants the leaves are greatly reduced and usually non-photosynthetic. They contain a single, non-branching vascular trace, which is the defining feature of microphylls. However, it has recently been recognised that these microphylls probably evolved by the reduction of a megaphyll; therefore they are commonly referred to as megaphylls to reflect this homology.

They grow in whorls fused into nodal sheaths. The stems are green and photosynthetic, also distinctive in being hollow, jointed, and ridged (with (3-) 6-40 ridges). There may or may not be whorls of branches at the nodes; when present, these branches are identical to the main stem except smaller.

Spores

The spores are borne under sporangiophores in cone-like structures (*strobilus*, pl. *strobili*) at the tips of some of the stems. In many species the cone-bearing stems are unbranched, and in some (e.g. *E. arvense*) they are non-photosynthetic, produced early in spring separately from photosynthetic sterile stems. In some other species (e.g. *E. palustre*) they are very similar to sterile stems, photosynthetic and with whorls of branches.

Horsetails are mostly homosporous, though in *E. arvense*, smaller spores give rise to male prothalli. The spores have four elaters that act as moisture-sensitive springs, assisting spore dispersal after the sporangia have split open longitudinally.

Habitat

Many plants in this genus prefer wet sandy soils, though some are aquatic and others adapted to wet clay soils. One horsetail, *E. arvense,* can be a nuisance weed because it readily regrows after being pulled out. The stalks arise from rhizomes that are deep underground and almost impossible to dig out. It is also unaffected by many herbicides designed to kill seed plants. The foliage of some species is poisonous to grazing animals if eaten in large quantities. *Equisetum* is cooked and eaten in Japan.

Geological History

The horsetails are the sole surviving genus of the Equisetopsida, a diverse and widespread group during the Carboniferous period. Some species were large trees reaching to 30 m[*verification needed*] tall. The genus *Calamites* (family Calamitaceae) is abundant in coal deposits from the Carboniferous period.

Species

Microscopic view of *Equisetum hyemale'*. 2-1-0-1-2 is one millimeter with 1/20th graduation. White and small protuberances are accumulated silicic acid on cells. Boiled and then dried *E. hyemale* is still used in Japan for the final polishing process on woodcraft to produce a smoother finish than with sandpaper.

Subgenus *Equisetum*

- *Equisetum arvense* - Field or Common Horsetail
- *Equisetum bogotense* - Andean Horsetail
- *Equisetum diffusum* - Himalayan Horsetail
- *Equisetum fluviatile* - Water Horsetail
- *Equisetum palustre* - Marsh Horsetail
- *Equisetum pratense* - Shade or Meadow Horsetail
- *Equisetum sylvaticum* - Wood Horsetail
- *Equisetum telmateia* - Great Horsetail

Subgenus *Hippochaete*

- *Equisetum giganteum* - Giant Horsetail
- *Equisetum myriochaetum* - Mexican Giant Horsetail
- *Equisetum hyemale* - Rough Horsetail
- *Equisetum laevigatum* - Smooth Horsetail
- *Equisetum ramosissimum* - Branched Horsetail
- *Equisetum scirpoides* - Dwarf Horsetail
- *Equisetum variegatum* - Variegated Horsetail

Named Hybrids

Hybrids between species in subgenus *Equisetum*

- *Equisetum* × *litorale* Kühlew ex Rupr. = *Equisetum fluviatile* × *Equisetum arvense*
- *Equisetum* × *dycei* C.N.Page = *Equisetum fluviatile* × *Equisetum palustre*
- *Equisetum* × *willmotii* C.N.Page = *Equisetum fluviatile* × *Equisetum telmateia*
- *Equisetum* × *rothmaleri* C.N.Page = *Equisetum arvense* × *Equisetum palustre*
- *Equisetum* × *robertsii* Dines = *Equisetum arvense* × *Equisetum telmateia*
- *Equisetum* × *mildeanum* Rothm. = *Equisetum pratense* × *Equisetum sylvaticum*
- *Equisetum* × *bowmanii* C.N.Page = *Equisetum sylvaticum* × *Equisetum telmateia*
- *Equisetum* × *font–queri* Rothm. = *Equisetum palustre* × *Equisetum telmateia*

Hybrids between species in subgenus *Hippochaete*

- *Equisetum* × *moorei* Newman = *Equisetum hyemale* × *Equisetum ramosissimum*
- *Equisetum* × *trachydon* A.Braun = *Equisetum hyemale* × *Equisetum variegatum*
- *Equisetum* × *schaffneri* Milde = *Equisetum giganteum* × *Equisetum myriochaetum*

- *Equisetum × ferrissii* Clute = *Equisetum hyemale × Equisetum laevigatum*
- *Equisetum × nelsonii* (A.A.Eat.) Schaffn. = *Equisetum laevigatum × Equisetum variegatum*

The superficially similar flowering plant, Mare's tail (*Hippuris vulgaris*), unrelated to the genus *Equisetum*, is occasionally misidentified and misnamed as a horsetail.

List of Plants Poisonous to Equines

This is a list of plants which are poisonous to equines. Some may cause mild reactions, such as diarrhea, others can lead to serious problems including horse colic, laminitis, and neurological problems, which, in some circumstances, can be fatal.

Division Magnoliophyta

Class Magnoliopsida

- *Acer rubrum*
- *Castor oil plant*
- *Dendrocnide moroides* (Stinging tree or Gympie stinger)
- *Cicuta virosa*
- Echium plantagineum (Paterson's Curse)
- Green Cestrum (C. parqui)
- Kalanchoe delagoensis (Mother of millions)
- Oak
- Oleander
- Pokeweed
- Quince (Cydonia oblonga)
- *Ranunculus*
- St John's wort
- *Nicotiana*

Order Asterales

- *Achillea ptarmica*
- *Artemisia tridentata*
- *Centaurea solstitialis*
- Cocklebur
- Ragwort
- White Snakeroot

Order Ericales

- *Kalmia latifolia*
- *Rhododendron* (including Azalea)

Order Fabales

- Black locust
- Locoweed

Order Lamiales

- *Digitalis*
- Privet

Order Laurales

- *Persea americana* (Avocado)

Order Solanales

- Solanaceae (including Deadly nightshade and Potato)

Class Liliopsida

- Johnson grass

Division *Pinophyta*

- *Taxus sp* (Yew)

Division *Pteridophyta*

- Bracken fern (also known as brake fern)
- *Equisetum* (Horsetail)

11

Plants without Flowers

INTRODUCTION

The little plants which live on leaves are called epiphylls. 'Epi' means 'upon' and 'phyll' means 'leaf'. (Epiphyte, the more familiar name used for a number of ferns, orchids and so on, means 'upon-plants'.) piphylls include a variety of non-flowering plants including lichens (the aingroup), mosses, leafy liverworts and algae. Twenty species, or more, may be found on a single leaf. They flourish best where light levels are low, on leaves in the understorey. Indeed, almost every understorey leaf supports at least a few of these plants and among them lives an entire microscopic fauna population of mites, worms and insect larvae as well as bacteria and a number of fungi species. We often talk of biodiversity in rainforests in terms of the bewildering number of tree species but the variety of microscopic epiphylls and their dependent animals make a very valuable contribution to this.

Many of the bacteria which live onleaves play an important part in the rainforest ecosystem by using nitrogen from the atmosphere and cycling nutrients. However, epiphylls can be quite a burden for a plant. A heavy population of them on a leaf significantly reduces its ability to photosynthesise (produce food from sunlight). To discourage this any rainforest leaves are shiny and their shape, designed to encourage water to run off quickly, is thought to reduce the ability of epiphylls to colonise them.

Nonetheless, epiphylls are an important part of the whole rainforest ecosystem. They are also a sensitive bunch and can be disturbed by human activities. Loss of habitat has left them vulnerable and recent research suggests that some may function as good indicators of forest stress.

Algae were the first proper plants to evolve and green algae are the ancestors of all of today's land plants. They are very simple cellular plants, with no roots, stems or flowers.

While most are found in water bodies, salt and fresh, the notorious luegreen algae forming blooms where nutrient loads are high, others are found on land where conditions are moist. Small forms grow as parasites below the 'skin' (the cuticle) of rainforest leaves.

Mosses represent an evolutionary step up from algae. They have no roots and no system of woody vessels which allow more advanced plants to grow tall. As the first land plants mosses probably created the first forests, mini-ecosystems just 5cm or so high, achieving this impressive height simply by packing themselves tightly together. The sex life of mosses resembles the complicated system of alternate generations found in algae.

However, mosses keep the next generation at home! Whereas algal male and female cells meet by wimming freely through the water, female moss cells are attached firmly to the parent plant. When they have been fertilised by free-swimming male cells they grow into spore-filled capsules on the end of long stalks.

These capsules eventually open to release the spores which are blown away to grow into new moss plants. In some respects, this tendency to retain the female egg on the parent plant resembles the habit of later plants, such as cycads, conifers and flowering plants, to do the same. Perhaps it represents a step along the evolutionary way.

LIVERWORTS

Liverworts are very similar to mosses. However, where the pointed 'leaves' of mosses are arranged in spirals on the stems, liverworts have rounded (liver-shaped) 'leaves' which grow flat in double rows.

Lichens could perhaps be regarded as terrestrial algae.

They are remarkable organisms. Each is a combination of an alga (blue-green or green) and a fungus, functioning together in a symbiotic partnership. The fungus provides the 'root system' (hyphae), drawing up water and minerals for the pair. The rootless alga has the chlorophyll, the green stuff in most plants which enables them to otosynthesise — to create food from sunlight. To do so the alga needs thewater provided by the fungus, while the fungus needs the food anufactured by the alga. Together they are a lichen.

Generally the fungal threads create the body of the lichen with the algal cells living within them. This gives the lichen strength and prevents the alga from drying out. The fungi also manufacture a type of acid which serves to eat away at underlying rock and provide the lichen with a foothold. Lichens therefore erode rock, very slowly. It has been estimated that they produce a centimetre of topsoil in two thousand years.

Many lichens produce small powdery granules called soredia on their surface. These contain both algal cells and fungal threads. When washed or blown away, or carried by insects, they grow into new lichens. Larger upright packets (isidia) work in a similar way. However, sometimes the fungal partners make their own fruiting bodies. These vary in appearance, resembling small volcanoes, cups or jam tarts. When these burst, the fungal spores are dispersed but if they do not pair up with the correct algal cells they cannot develop. (The algal cells, on the other hand, are able to live independently.)

Lichens grow on rocks, walls, trees, fences, roofs and on the ground, as well as on leaves. They are limited by light, needed by the algae for photosynthesis, and lso need clean air. The presence of sulphur dioxide, produced when fuels such as coal and oil are burned, and which is esponsible for acid rain damage, kills lichens. These plants are therefore good indicators of a clean environment. Since some are more tolerant of pollution than others, the type of lichen growing can tell us how bad he pollution is. However, lichens are very hardy in natural conditions, living in the tropics, on mountain tops, in salty seashore splash zones nd even polar regions, where they survive temperatures well below reezing. They can stop growing if conditions are dry, resuming when moisture returns and are usually the first organisms to invade barren ground, such as new lava flows. Here they provide a foothold for other plants to move in.

FERNS — THE ADVANCED MODEL

In the evolutionary development of plants, ferns represent a great ance on all previous models. Whereas all the surface cells of an aquatic alga are able to absorb water and nutrients, on land it is ecessary to divide up the tasks, a step which makes the plants more adaptable.

- Roots were a revolutionary new feature dedicated to seeking out less accessible sources of water, thus allowing plants to move inland. They also served to stabilise the larger models.
- Water and nutrients, taken up by the roots, had to reach other parts of the plant so a plumbing system — the vascular system — evolved. Woody vessels (xylem) performed this function, moving water and utrients upwards. These vessels had a duel function, providing rigidity to the tissues. With these load-bearing structures the plants were able to grow much taller and reach up to the light.
- Leaves were another fern invention — a system of solar panels edicated to capturing the energy of the sun and turning it into food. Exposed to the air, these had to be sealed to prevent the water gath ed by the roots from leaking away so a waxy skin (cuticle) was developed. Since the process of photosynthesis requires an intake of carbon ioxide from the atmosphere and waste oxygen must be released, special design features in the cuticle — pores — allowed this exchange of gases to continue.
- Another plumbing system was needed to move the sugars and other photosynthetic products from the leaves to the rest of the plant. This function was performed by another new system of vessels, the phloem.

Although ferns were among the earliest vascular plants (algae, lichens, mosses and liverworts are all classed as non-vascular plants) they were not the only ones. The fossil records tell us that at one time the world was dominated by massive clubmosses, giant horsetails and others hich created magnificent forests 45m or more in height as they used ir newly developed vascular systems to reach higher and higher in competition for sunlight. Many of these plants are now extinct, their relatives hanging on in comparative obscurity.

WATER FERNS

Certain ferns live in water. Some have their roots in the mud while others are free floating. Nardoo (*Marsilea* species, left) is an unusual fern with leaves hich look like four-leaved clovers. It grows in swamps and still lakesides with its roots and creeping stems in the mud and its leaflets floating on the water surface. (The early explorers, Burke and Wills, died from starvation while feeding on the starchy parts of this fern. It is thought that chemicals in it interfere with our ability to absorb nutrients.) The dreadful exotic weed from Brazil, *Salvinia molesta*, (right) is a eefloating fern. Groups of three leaves grow along a thin stem. Two of the leaves contain large air spaces and act as floats while the third is divided into threadlike hairy lobes which dangle in the water acting as roots to absorb water and nutrients. This fern multiplies vegetatively with great speed, covering water surfaces and cutting out light and air to other aquatic organisms.

Coral Ferns

Coral ferns are scrambling ferns which form tangled masses. The multi-forked fronds, which form almost geometrical patterns, can grow to 4m on a long, brown, wooly stalk. They are found in large colonies in a variety of wet sunny sites. They like to have their roots in water and their fronds in the sun.

Filmy Ferns

These are extremely delicate ferns with leaf blades just one cell thick. They grow best in humid tropical conditions because the entire leaf surface is open to the atmosphere and they can dry out very easily.

Maidenhair Ferns

Popular with gardeners and as house plants, maidenhair ferns (left) usually have fine black stems and frilly fanshaped leaflets. Some species are very common in the Wet Tropics, growing in moist open sites along river and track banks.

TREE FERNS

There are seven species of tree ferns in the Wet Tropics, three of them endemic. Although all seven are found in the uplands, only two, the caly, or Cooper's, tree fern (*Cyathea cooperi*) and black tree fern *athea rebeccae*) are found in the tropical lowlands. The most ommon and most widespread, the scaly tree fern, is also the largest, rowing to a height of 15m with a 30cm thick trunk. The trunk atterned with large oval scars left by fallen fronds. It is hardy and often grown in ardens. Unlike conifers and other advanced trees, tree ferns are not capable of producing secondary timber. This means that their trunks cannot become thicker, growth occurring only from the top. The trunk is pithy on the inside but hard on the outside and is often covered with a mass of tough aerial roots. If the tree fern is knocked over these can grow into the ground and develop as normal roots.

Epiphytic Ferns

Many ferns types grow on other plants, using them as perches but not stealing any nutrients from them. A number of ferns specialise in this lifestyle in tropical rainforests because the forest floor is too gloomy for the to survive and their only alternative is to perch on the trunk or branches of a tree. They can also survive on rocks and on the ground if there is sufficient light but away from the forest edge it is the spores which land high in a tree which flourish. Brown sterile leaves hold basket, elkhorn and staghorn ferns in place while the green leaves photosynthesise and produce spores. Trapped dead leaves and other debris provide nutrients.

Hare's Foot Fern

This fern (below) is named for the brown rhizome, or stem, covered with brown papery scales which may protrude from the clump, above the ground, for up to 50cm. It tends to be an epiphyte, growing on rainforest trees, on rocks or on the ground. Aborigines have traditionally boiled the roots and stems to treat morrhaging. King fern (*Angiopteris evecta*) Easily mistaken for a trunkless palm, the king fern produces possibly the longest fern fronds in the world, growing up to 7m in length. Fronds usually sprout from near ground level, trunks (butts) developing slowly. King ferns like dimly-lit rainforest stream banks and plenty of water. The related potato fern (*Marattia oreades*) has weeping fronds up to 2m long. Like tree ferns both these giant ferns have an ancient history. Fossils well over 300 million years old, and very similar to the modern versions, have been found on most continents. They predate the dinosaurs.

Fern Allies

A 420 million-year-old clubmoss fossil found in Australia is one of the oldest known land-plants. Although they once grew as enormous trees, present-day clubmosses are more modest plants, known as fern allies. They are not mosses but the arrangement of their tiny pointed leaves in spirals around the stems is moss-like in appearance. They are credited with producing the first cones — long before the conifers developed them to perfection. These are formed from tightly packed leaf-like structures each of which bears spores beneath it. Fern allies produce two types of spore, small ones which produce male thalli and large ones which produce female thalli.

The sperm must then swim the extra distance to find a female thallus before fertilisation can take place. Selaginella species (left) are pretty, multi-branched fern allies which form low dense carpets in damp shady areas. They too have an ancient history and once formed forests 40m high. Now they rarely reach more than 10cm in height and scramble along the ground producing new roots at intervals along the stems. Little scale-like leaves line the stems. Thin, elongated tips on some branches are the spore-bearing cones made up of modified leaves.

Tassel Ferns

This an ancient group of fern allies which has declined dramatically since their peak in the Carboniferous era. The rock tassel fern (*Huperzia squarrosa,* right) which grows in the Wet Tropics today is very similar to 415 million-year-old fossils from Victoria. Tassel ferns are epiphytes, with long dangling stems. Some end in long tassel-like 'clubs' which are the spore-bearing cones, rather like those on Selaginella. However, in *H. squarrosa* the fertile spore-bearing leaves are the same as the normal leaves, a characteristic considered very primitive and indicative of its ancient lineage. Related to the tassel ferns is a vigorous scrambling ground-creeping version (*Lycopodium cernuum*) which is common along sunny roadsides. It looks like a tiny pine tree with little cones on the erect branches.

12

Filicophyta

INTRODUCTION

The ferns are vegetable vascular. They constitute the subdivision of the *Pterophyta* (or *Pteridophyta Strictly speaking*), which counts approximately 13 000 species (the largest vegetable junction after the Angiosperme S). One meets approximately the three-quarters of the species in the tropical areas and a good proportion of these tropical ferns is épiphyte.

The division (paraphyletic) of the Ptéridophyte S (*Pteridophyta*) includes/understands, in addition to the filicophytes themselves, the Șilophyte S (*Silophyta*) which is brought today closer to the Ophioglossophyte S (*Ophioglossophyta*, to see low), but also the Sphénophyte S (*Sphenophyta*, of which prêles or *Equisétophyta*) and the Lycophyte S (*Lycophyta*, of which lycopods).

The majority of the ferns, leptosporangié be, are members of the clade *Filicophyta*. The others are members of other clades. Morphologiquement, the ferns have a rather great diversity. Certain tree ferns exceed 20 m in height but them Stipe does not present secondary growth in thickness. The well developed sheets are often made up mégaphylles with circinée prefoliation (in stick).

VEGETATIVE CYCLE

- the tendency to the development of the phase Diploïde (sporophyte) is marked. The Gamétophyte is a generation independent in the shape of a plant of small size called Prothalle.
- the majority of the current ferns are isosporous, i.e. the Sporophyte gives only one kind of spores which after germination produces a gamétophyte carrying Archégone S and Anthéridies.
- the Sporange S grouped in clusters called sores protected by Indusie S can have localizations différentes:
  - On the margin or the lower face of the sheets.
  - On modified sheets.
  - On distinct branches.

Classification of the Ferns

The study of the formation of the sporanges results in classifying the ferns in two great groups:

- the ferns eusporangié be:
  - Marattiales
  - Ophioglossales (of which Ophioglossophyte S)
- the ferns leptosporangié are:
  - the ferns isosporous are:
    - Polypodiales
  - the ferns hétérosporé are:
    - Hydropteridales

The phylogenetic relations between the various families of ferns still make the subject of debate and following classification is given as example:

- *Marattiidae*
  - order *Marattiales*
  - family *Marattiaceae*

- *Ophioglossidae*
  - order *Ophioglossales*
  - family *Ophioglossaceae*
- *Polypodiidae*
  - order *Hymenophyllales*
  - family *Hymenophyllaceae*
  - family *Hymenophyllopsidaceae*
- order *Osmundales*
  - family *Osmundaceae*
- order *Gleicheniales*
  - family *Gleicheniaceae*
  - family *Platyzomataceae*
  - family *Matoniaceae*
  - family *Stromatopteridaceae*
  - family *Dipteridaceae*
  - family *Cheiropleuriaceae*
- order *Schizaeales*
  - family *Schizaeaceae*
  - family *Lygodiaceae*
- order *Plagiogyriales*
  - family *Plagiogyriaceae*
- order *Dicksoniales*
  - family *Dicksoniaceae*
  - family *Dennstaedtiaceae*
- order *Cyatheales*
  - family *Cyatheaceae*
  - family *Metaxyaceae*
  - family *Loxsomataceae*

- order *Marsileales*
  - family *Marsileaceae*
- order *Salviniales*
  - family *Salviniaceae*
  - family *Azollaceae*
- order *Pteridales*
  - family *Actinopteridaceae*
  - family *Adiantaceae*
  - family *Parkeriaceae*
  - family *Pteridaceae*
  - family *Vittariaceae*
- order *Blechnales*
  - family *Blechnaceae*
  - family *Aspleniaceae*
  - family *Dryopteridaceae*
  - family *Lomariopsidaceae*
  - family *Thelypteridaceae*
  - family *Woodsiaceae*
- order *Davalliales*
  - family *Davalliaceae*
  - family *Oleandraceae*
- order *Polypodiales*
  - family *Polypodiaceae*
  - family *Grammitidaceae*

The recent studies resulted separating Ophioglossales from the other ferns and in creating the junction of the Ophioglossophyta.

EUKARYOTA (PHYLOGENETIC CLASSIFICATION)

This chapter has the aim of presenting a Phylogenetic tree Eukaryota, i.e. a tree clarifying the complex evolutionary relations (relations ancestral groups - downward groups) existing between the various living organisms, or taxed, component this group. Because of its new methods based on molecular biology, the Phylogeny, recent science at the origin of the phylogenetic Classification of alive, is in full rise and sometimes in contradiction with old the traditional Classification, mainly based, it, on morphological and physiological criteria. Also, certain particular points are sometimes still the debate object within the scientific community. The objective, if it is necessary, is also to present the latter ici.

This tree presents the still existing organizations as well today as the extinct groups (including the fossil groups).

PHYLOGENETIC TREE

The sign **(+)** returns to the phylogenetic classification of the group considered.

Summarized Tree

O LUCA ¦ +-o Eubacteria (ex- Bacteria) +-o +-o Archaea +-o EUKARYOTA +-o Unikonta ¦ ¦ ¦ +-o Amoebozoa (+) ¦ ¦ ¦ +-o Opisthokonta ¦ ¦ ¦ +-o Fungi or MYCOTA (+) ¦ ¦ ¦ +-o Mesomycetozoa (+) ¦ ¦ ¦ +-o METAZOA (+) ¦ +-o Bikonta +-o Apusozoa +-o +-o Cabozoa ¦ ¦ ¦ +-o Excavata (+) ¦ ¦ ¦ +-o Rhizaria (+)) +-o +-o Chromalveolata ¦ ¦ ¦ +-o Chromista (+) ¦ ¦ ¦ +-o Alveolata *(Sil)* ¦ +-o ¦ ¦ +-o Dinoflagellata (+) ¦ ¦ +-o Apicomplexa (+) ¦ +-o Ciliata (+) ¦ +-o Plantae or ARCHAEPLASTIDA (+)

Developed Tree

O LUCA ¦ +-o Eubacteria (ex- Bacteria) +-o +-o Archaea +-o Eukaryota +-o Unikonta ¦ ¦ ¦ +-o Amoebozoa (+) ¦ ¦ ¦ +-o Opisthokonta ¦ ¦ ¦ +-o Fungi or MYCOTA (+) ¦ ¦ ¦ +-o Mesomycetozoa (+) ¦ ¦ ¦ +-o METAZOA (+) ¦ +-o Bikonta +-? Apusozoa ¦ +-o Diphyllatea or Collodictyonidae ¦ +-o Thecomonadea ¦ +-? Hemimastigophora or Spironemidae ¦ +-o

Ancyromonadidae ¦ +-o Apusomonadida +-o +-o Cabozoa ¦ ¦ ¦ +-o Excavata (+) ¦ ¦ ¦ +-o Rhizaria (+) ¦ +-o Corticata +-? Centrohelea or Centroheliozoa ¦ +-o Axoplasthelida or Gymnosphaeridae ¦ +-o Centroplasthelida ¦ +-o Heterophryidae ¦ +-o Acanthocystidae ¦ +-o Raphidiophryidae +-o Chromalveolata ¦ ¦ ¦ +-o Chromista (+) ¦ ¦ ¦ +-o Alveolata or Alveolobionta ¦ +-o ¦ ¦ +-o Ellobiopsidae ¦ ¦ +-o Colponemea ¦ ¦ +-o Miozoa or Myzozoa ¦ ¦ +-? Apicomonadea or Colpodellidae ¦ ¦ +-o Dinozoa ¦ ¦ ¦ +-o Perkinsea ¦ ¦ ¦ ¦ ¦ ¦ ¦ +-o Dinoflagellata (+) ¦ ¦ ¦ ¦ ¦ +-o Apicomplexa (+) ¦ ¦ ¦ +-o Ciliophora or Ciliata (+) ¦ +-o Plantae or ARCHAEPLASTIDA (+)

| valign=" top" width=" 3%" | | valign=" top" width=" 25%" | |}

PHYLOGENETIC CLASSIFICATION

The phylogenetic classification is a system of classification Systématique of the living beings. It tries to replace the scientific Classification traditional based on multiple features: biological, phenotypical (anatomical) and physiological (physicochemical phenomena, nutrition). One of the characteristics of the phylogenetic approach is that this classification upsets classifications creationists like that developed by Carl von Linné. The binomial nomenclature of Linné was based on the proverb that all the species appeared at the same time and that those were fixed. Whereas phylogenetic classification illustrates the principles of evolutions and relationship of the species.

Presentation

Willi Hennig is at the origin of phylogenetic classification in 1950, it revolutionized thereafter, after the translation in English of its work, all the Systématique starting from the end of the Années 1960. The anatomical characters of the living beings then were analyzed differently, joined soon by the molecular characters. Phylogenetic classification is a system of classification of the living beings initially founded on what the living beings have (at the morphological level as at the molecular level), and not, in first authority, on what they do not have (one will avoid then "invertebrate") or on what they make (one will avoid then "viviparous", "digger"). It is not an ecological classification nor anthropocentric. It gathers the living beings on a particular type of resemblance, those of the resemblances which are évolutivement innovating within the sample of species to classify. She is opposed in that to classifications Phénétique S which do not sort the resemblance and make classifications starting from the total resemblance. She replaces the traditional classification based on a crowd of criteria, morphological but so ecological, ethologic, anthropocentric, and preferences nutritional.

Phylogenetic classification validates only groups Monophylétique S, whatever the characters used, molecular, morphological, anatomical, caryologic, genomic. The characters used to build a Phylogenetic tree are valid only if those are characters Apomorphe S shared by at least two Taxon S. (a character apomorphe is an ancestral character which was transformed. This character apomorphe divided is a Synapomorphie. Only the synapomorphies (or characters derived) characterize a Clade or groups monophyletic. A group Monophylétique (or clade) is a group including/understanding an ancestor (or node) and all his descendants.

The first hierarchy is thus made up of three fields:

Archée S

Bacterium S

Eucaryote S

To know which of these three groups divide a common ancestor which distinguishes them from the third is a subject of research.

Systematic the phylogenetic in its modern version rejects any categorization of the hierarchical levels. For practical reasons, the Arbre of alive the gives itself the hierarchy which the old categories tried to provide. The tree of alive is, indeed, a whole of branchpoints. Each level is compared to a Taxon. A taxon is a regrouping of organization defined by a whole of synapomorphies. The points or nodes are obligatorily a theoretical organization which would have the synapomorphies posterior nodes in the cladogramme. If this organization is discovered (fossils or others) a new tree must be built. In the long term, if the tree restores the totality of the relations of relationship, all the connections should be binary. The direction of the dichotomies has nothing to do with sexuality nor with binary speciations, because the tree is not a genealogy (" who goes down from qui"), but reflects only one explanatory capacity maximum of the tree: a completely dichotomic tree succeeds in restoring all the relations of relationship ("who is closer to qui").

The systematic modern one takes into account all the characters héritables, since what is visible (anatomy and morphology, base traditional classification) until the sequences of DNA and ARN, while passing by the Protéine S and the data of the Paléontologie. The sequencing of certain parts of the Genome, like DNA of the Mitochondries or ARN of the Ribosomes made it possible in the last years to make important progress in classification and to solve many secular problems.

The tree is obtained starting from many criteria and is called upon complex algorithms. Different algorithms can give different results. In this case, that which will answer more the criteria of parsimonies will be retained. I.e. the tree asking for the minimum of transformation of characters. Current classification is continuously altered according to new information. The site treebase refers. The examples which follow can have to be modified.

Example of the Man

A detailed example makes it possible to have an idea of the difference in result, compared to the traditional approach. The example impossible to circumvent is that of the Man. Here part of the successive nodes allowing to classify the man (certain intermediaries were omitted arbitrarily) such as appearing in *phylogenetic Classification of alive the* (Guillaume Lecointre and Herve Guyader, Belin edition) (description associated with each clade is not very rigorous and just aims at fixing the ideas):

- Species *Homo sapiens*
- Kind *Homo:* includes the fossil species like *Néanderthal* or *erectus*

- Hominines: the kinds *Australopithecus* and *Homo*
- Homininés: two species of chimpanzees and the hominines
- Family Hominides: gorilla and Homininés
- Hominoïdés: one adds the orang-outang
- Hominoïdes: one adds gibbons them
- Catarrhiniens: monkeys of the old world
- Simiiformes: Monkey S
- Haplorrhiniens: monkeys and Tarsier S
- Order Primate S (*strictly speaking*): most of the old order of the primates
- Primate S (*lato sensu*): the Scandentiens (toupayes) are added
- Archontes: are added the Chiroptère S (bat) and the Dermoptère S
- Preptothériens: are added the majority of the placental mammals
- Euthériens: the pangolins and fourmiliers (Xénarthres) are added to supplement the placental mammals
- Thériens: the marsupials
- Classe Mammifère S are added: the old class of the same name, precedents plus the Monotrème S (ornithorhynques and echidnas)
- Amniote S: the old class of the reptiles and the birds
- Tétrapode S are added: approximately the Amphibians
- Sarcoptérygien S are added: are added Dipneuste S then Coelacanthe S
- Ostéichthyen S: the actinoptérygiens are added (the major part of "osseous fish")
- Vertébré S: the Requin S then the Lamproie S
- Craniates are added: the Myxine S

- Embranchement Chordé S are added: amphioxus and ascidies
- Deutérostomien S are added: Echinodermata are added (starfishes, sea urchins,...)
- Bilatériens: symmetrical animals: are added to the precedents, inter alia, the protostomiens which include, between many others, various groups of worms, the arthropods (insects), the molluscs (snails)
- Eumétazoaires: organized animals, are added to the precedents the cnidaires for example
- Métazoaires: the old animal kingdom, are added to the precedents various groups of sponges
- Opisthocontes: are added, inter alia, the mushrooms
- Domaine Eucaryote S: living beings with cells with core; are added to the precedents listed above the green line (green algas, red algas, plants with flowers), the brown line (brown algas,...) and a whole collection of unicellular groupings of species.

Differences with Traditional Classification

Let us quote finally some examples of spectacular changes compared to the traditional classification.

- All the ancestors of the Dinosaure S are also ancestors of the Oiseau X, which brings to regard those as the descendants of a group of small alive flying dinosaurs to the Crétacé. The concept of "Reptile" is abandoned.

Let us specify the relation between birds and dinosaurs. For certain authors, the Vélociraptor is regarded as nearer to the birds than the Tyrannosaure, and this last is considered nearer to the birds than Diplodocus.

- more the close relatives of the Cétacé S would be the Hippopotame S. the concept of "Artiodactyle S" is abandoned.
- more the alive close relatives of the birds are the Crocodile S.

- the Poisson S osseous are closer to the mammals than Requin S. the concept of "fish" is abandoned.
- the old group of the "Algue S" explodes in all directions, some being grouped with the house plants, others with the Bactérie S.
- the concept of "Protiste" is abandoned, with the profit of groups which can mix multicellular beings and monocellulaires (for example the Straménopile S gather the brown Algue S - of which the Kelp, up to 60 m of length - and the unicellular Diatomée S).
- Procaryotic division between S and Eucaryote S is abandoned, with the profit of division into three of the alive one.
- the application of this classification to the Angiosperme S is illustrated by the Classification APG (Angiosperms Phylogeny Group).

PHYLOGENY

The phylogeny is the study of the formation and the evolution of the living organisms in order to establish their relationship. The phylogenesis is the term more used to describe the Généalogie of a Espèce, of a group of species but also, on an intraspecific level, the genealogy between Population S or individuals.

One usually represents a phylogeny by a Phylogenetic tree. The proximity of the branches of this tree represents the degree of relationship between the Taxon S, the nodes the common ancestors of the taxons. In a tree worked out by Phénétique, the length of the branches represents the genetic distance enters tax; in a tree worked out by Cladistique (cladogramme), one places on the branches the evolutionary events (derived characters) having taken place in each line.

Presentation

The Systematic, the study of biological diversity for its phylogenetic Classification, concentrates, in the light of the recent discoveries, on a phylogenetic Classification replacing the traditional Classification now. Traditional classification establishes groups or tax according to a simple criterion with total resemblance. A phylogenetic classification supposes that one gathers the living beings according to their family ties. Any systematic group (or "taxon") thus contains close living beings between them genetically (what is not always correlated with a phenotypical resemblance total). The family ties between two members of a taxon are increasingly closer than the family ties between an unspecified member of the group and an living being external with the group (it happens that this external member is however very resembling because of the phenomenon of evolutionary Convergence). To reconstitute the family ties between living beings, phylogeny proceeds according to two techniques: the phenetic and the cladistic. The phénétique one.

The phénétique one rests on the basic postulate that the degree of resemblance is correlated with the degree of relationship. It thus supposes to quantify the resemblance between the living beings to be classified.

This method appears not very relevant when one applies it to the morphological characters because of the Analogie S: certain resemblances between living beings or let us tax cannot indeed be allotted to a common ascent. One speaks then about analogy. The principle used to explain this phenomenon is evolutionary convergence: two let us tax different living in ecological Niches similar or on which the natural selection had a similar impact could be similar. The wings of the birds and the bats are similar characters as wings, because these two wings are not inherited a winged common ancestor. Moreover it is very difficult to quantify morphological resemblances numerically.

On the other hand, the phénétique one becomes relevant since one compares a very great number (with the statistical direction) of characters because the number of similar natures becomes negligible among all the characters whose resemblance is indeed due to the relationship. Thus this technique is very powerful when one applies it at the molecular level. The systematicians have thus more and more recourse to molecular methods to compare tax them and to rebuild phylogenies. With this intention, they compare various molecular components from alive like DNA, ARN or the Protéine S. Indeed, DNA, ARN and proteins are molecules Polymère S. Each residue of the molecule (Nucléotide for the DNA and the ARN or Amino-acid for protein) can be regarded as a character. It is thus possible to compare the sequences at several living beings and to quantify their resemblance by a simple percentage that one compares to the genetic distance between the two let us tax to which the two living beings belong. The results are represented in a Phylogenetic tree, which one could name phénogramme, where the length of the branches depends on the genetic distance and thus represents the degree of relationship between tax studied.

The Cladistique One

The cladistique one initiated by Hennig treats on a hierarchical basis the compared characters. In fact are gathered in same a taxon only the living beings which share characters homologous S: when a resemblance between two tax can be allotted to a common ascent, one speaks about homology. The forelimbs of all the Tétrapode S, which they are arm or wings, are homologous.

Thus the wing of the bat and the bird are homologous as forelimbs, and not as wings. The common ancestor of the bird and the bald person mouse had indeed already four legs but its forelimbs were not wings. This common ancestor is indeed also that of the Lézard S, of the Crocodilien S. the forelimb "wing" appeared later independently in the two lines Chiroptère S and Oiseau X...

The homologies in fact are seen like shared evolutionary innovations (Synapomorphie S): if the same homologous character is shared by two let us tax it is that both tax inherited it their common ancestor. This homologous character thus appeared in the line leading to this common ancestor. Any living being having this homologous character thus goes down from this common ancestor. Any living being not having this homologous character does not go down from this common ancestor and is thus distant genetically.

The cladistique one thus rests on the identification (often difficult) of the homology of the characters. It is relevant at the morphological level (and is thus the only means of classifying the fossil species whose DNA is seldom preserved) as at the molecular level. The results are represented in a phylogenetic tree or cladogramme in which each node represents a common ancestor and where the synapomorphies are represented on the branches of which the length is arbitrary. Two let us tax are all the more related as they divide a common ancestor near in the tree. Here also, therefore, let us tax them find themselves gathered according to their family ties.

Joint use of Phenetic and the Cladistique One

For a long time sometimes violent discussions opposed holding of one or other technique. Today the phénétique one and the cladistique one is often used jointly as being two independent methods. When their results are convergent, very solid phylogenies are obtained.

The joint use of these two methods revealed the existence in the traditional Classification of many groups nonfounded on the family ties and which are thus regarded as nonlegitimate and do not owe any more beings used in Taxonomie. The use of phenetic molecular and cladistic as well as confrontation of the trees obtained was largely allowed by the modern methods which are amplification by PCR and the Séquençage, allied with powerful computational tools which make it possible to automate these methods.

Example of changes in the phylogenetic tree due to the use of these techniques:

- the group of the reptiles. Were gathered within this one the Crocodilien S (makes some genetically close to the Oiseau X) and the Lézard S, Serpent S and tortoise S (distant genetically from the birds).

Example of the use of the Gene 16s for the studies of phylogeny of the Procaryotic.

13

Pteridosperms

INTRODUCTION

Pteridosperms or seed ferns are a very heterogeneous group of extinct plants with mostly fern-like foliage but with real seeds. They are mostly reconstructed as small trees but also forms with a climbing growth habit gave been found. Some forms, notably Medullosales, have really large fronds which could be up to 7 m long. Several groups can be distinguished within the pteridosperms. The classification of seed ferns is primarily based on fructifications and/or anatomical features. Eight groups of pteridosperms are presently recognised, six of them are known from the Palaeozoic and three from the Mesozoic. Some groups are very well known including their reproductive organs, whereas others are still very poorly understood and based on anatomically preserved vegetative remains, mostly axes.

PTERIDOSPERMS

Pteridosperms evolved in the latest Devonian (Fammenian) and became more common in the Carboniferous. The Lyginopteridales are most common in the Namurian and Lower Westphalian. The Medullosales took over the leading role during the Westphalian became less common in the latest Stephanian and persisted into the Permian. The small group of the Callistophytales is known from the Upper Westphalian to Lower

Permian. Peltaspermales evolved in the late Stephanian and were most common in the Triassic. The essentially Permian Glossopteridales are typical Gondwana elements. Corystospermales and Caytoniales are Mesozoic pteridosperms.

The Medullosales are one of the most widespread groups Palaeozoic seed ferns. They have rather large seeds which can be several cm long; one of the seed types is *Trigonocarpus*. Anatomically preserved seeds have been described as *Pachytesta*. The pollen organs are relatively complex, usually consisting of a (large) number of fused pollen sacs; one of the pollen organs has been described as *Bernaultia*. They produced large prepollen grains, e.g., *Schopfipollenites*. Fronds are typically bifurcated. Whittleseya is another type of pollen organ. Foliage form-genera like *Neuropteris* and *Alethopteris* are commonly attributed to the Medullosales. Medullosales had different growth forms, varying from medium-sized tree fern-like plants with very large fronds to liana-like plants with rather small compact fronds.

PTERIDOSPERM FOLIAGE

Pteridosperm foliage is often difficult to distinguish from fern foliage. In fact, the natural affinity of many Carboniferous foliage types is still unknown. Foliage is therefore classified in so-called form-genera; such form-genera of foliage types of unknown natural affinity (ferns + pteridosperms) are usually listed under pteridophylla. Form-genera are defined on the basis of pinnule outline and venation pattern. Most Carboniferous fern-like foliage appears to be seed fern foliage. Common foliage types are *Neuropteris*, *Alethopteris* and *Sphenopteris*. Many of these foliage taxa are valuable index fossils.

ALETHOPTERIS

Alethopteris is a form-genus for fern-like foliage with tongue-shaped, decurrent pinnules. The venation is pinnate with a strong, usually sunken midvein; some additional smaller veins arise directly from the rachis (in the decurrent basal part of the pinnule). Fonds are typically bifurcated and large to very large; however, most of the specimens found are rather small. *Alethopteris* is commonly considered to be medullosan foliage.

Neuropteris is another common foliage type. It is characterized by it tongue-shaped pedicellate pinnules which have a pinnate venation. *Neuropteris* fronds are usually bifurcated and large to very large. *Neuropteris* is also commonly considered to be medullosan foliage.

Sphenopteris and *Eusphenopteris* are two form-genera for foliage types with (strongly) dissected or lobed pinnules. Euspenopteris has pinnules with rounded lobes, whereas *Sphenopteris sensu stricto* has more dissected lobes. Euspheno-pterid fronds are considered to be pteridosperm foliage. However, *Sphenopteris* is a form-genus that is known to include seed ferns and true ferns. Lyginopteridales have bifurcted fronds with sphenopterid pinnules.

Mariopteris is a form genus for relatively small compact fronds which are bifurcated twice. The pinnules are tongue-shaped to assymmetrically triangular in outline with a distinct basal lobe (at least the pinnules closest to the main axes). Although it is clear that most mariopterids are seed ferns, their natural affinity remains unclear. Mariopterids obviously had a climbing growth habit.

Reticulopteris is a small genus including fronds with pinnules having a neuropterid shape but differing in having a reticulate venation pattern

Cyclopteris is a form-genus for large circular pinnules, up to ca. 10 cm in diameter, which are found in the basal parts of *Neuropteris* fronds or in the basal parts of its pinnae.

In contrast to true ferns Pteridosperms have well developed, very resistant cuticles. Each genus, or in most cases even every species, has its own typical epidermal pattern reflected in the overlying cuticle. Cuticles are therefore very useful in taxonomic. Moreover, cuticles also display a number of other features that can be used in palaeoecological and palaeoclimatological studies.

In addition to the taxa listed above there are several other (groups of) pteridosperms. In the Euramerian Upper Carboniferous and Lower Permian these include the Callistophytales and Peltaspermales.

Index